Synchronization and
Arbitration in Digital Systems

Synchronization and Arbitration in Digital Systems

David J. Kinniment
School of EECE, University of Newcastle, UK

John Wiley & Sons, Ltd

Other Wiley Editorial Offices

John Wiley & Sons Inc., 111 River Street, Hoboken, NJ 07030, USA

Jossey-Bass, 989 Market Street, San Francisco, CA 94103-1741, USA

Wiley-VCH Verlag GmbH, Boschstr. 12, D-69469 Weinheim, Germany

John Wiley & Sons Australia Ltd, 42 McDougall Street, Milton, Queensland 4064,
Australia

John Wiley & Sons (Asia) Pte Ltd, 2 Clementi Loop #02-01, Jin Xing Distripark, Singapore
129809

John Wiley & Sons Canada Ltd, 6045 Freemont Blvd, Mississauga, ONT, L5R 4J3

Wiley also publishes its books in a variety of electronic formats. Some content that appears
in print may not be available in electronic books.

Anniversary Logo Design: Richard J. Pacifico

British Library Cataloguing in Publication Data

A catalogue record for this book is available from the British Library

ISBN 978-0470-51082-7

Typeset in 10.5/13pt Sabon by Thomson Digital
Printed and bound in Great Britain by TJ International Ltd, Padstow, Cornwall
This book is printed on acid-free paper responsibly manufactured from sustainable forestry
in which at least two trees are planted for each one used for paper production.

Contents

Preface

Most books on digital design only briefly touch on the design of synchronizers and arbiters, with maybe two or three pages in a 300 page book, or a chapter at most. This is because there was no real need for it in the early years of computer design. Processors were largely self-contained and used a single clock, so interfacing the processor to slow peripherals, or other processors was not seen as a major task. The fact that it is not simple emerged in the 1970s and 1980s when data rates between processors increased, and sometimes systems with more than one time zone were being designed. Despite frequent synchronization failures because of lack of understanding of the design principles at that time, synchronization still did not make it into the standard literature, and very little has been written since about how they should be designed. More recently processors are being designed with many more high-speed ports linked to networks, and the systems themselves are often made up of several core processors connected to an internal bus or network on chip. This means that processors operating on different time frames must communicate at high data rates, and when two or more processors request access to a common resource, there has to be some arbitration to decide which request to deal with first.

The need for synchronizers to ensure that data coming from one time frame is readable in another, and arbiters to ensure that a clean decision is taken has always been there, but the understanding has not. Our aim is to promote good design in these areas because the number of timing interface circuits is escalating as the number of multiprocessor systems grows. A single processor interfacing to the real world through a few slow peripherals will not have many problems, but as the number of input/output ports increases, and the data rates increase, difficulties with reliability, data latency and design robustness will also increase.

This book has been written to meet the need for an understanding of the design of synchronizers and arbiters. It is intended for those involved in the design of digital hardware that has to interface to something else; other hardware, a communication system, or in the end, people. Only systems that do not interface to the outside world are free from the need to deal with synchronization and arbitration. It is divided into three sections. Section I deals with the fundamental problem of metastability. Any system that has to make a choice between two similar alternatives will end up taking longer and longer as the two alternatives approach each other in desirability, and if the amount of time available is limited, the decision mechanism will fail at a predictable rate. We describe the theory of this and how it affects practical circuits, so that the reader may be able to choose a suitable circuit for a particular application, and how to measure its reliability and performance. Section II looks at synchronizers in systems. In a multiprocessor system, the timing in each processor may be completely independent, linked by stoppable clocks as in globally asynchronous locally synchronous (GALS) systems, or partly linked for example by means of phase-locked loops. To optimize the system performance and reliability the synchronization method should be chosen to fit the methodology, and several examples are given. Arbitration has a section of its own, Section III, where the design of arbiters is approached by starting from a specification and developing asynchronous arbiters from simple daisy-chain circuits up to fully dynamic arbiters taking account of the priority required by each packet of data.

I am indebted to many of my colleagues for discussions over the years about the design of computer systems, but most particularly to those involved with the ASYNC series of seminars where interfacing systems using different timing methods, synchronous, asynchronous, and clocks with different frequencies has been a recurring theme. Because asynchronous systems are themselves made up of many high-speed components interacting together, it is there that the problems of timing are at their most acute, and many of the methods described had their first outing at one of the ASYNC seminars. Colleagues at Newcastle University have not only contributed to the text and reviewing of much of the manuscript, but have also provided much to the ideas contained in the book. In the end what matters is whether the techniques are applicable in the industrial world, and for this reason I am grateful for the input provided over many years from people who designed computers ranging from the Ferranti ATLAS in the early 1960s to SUN and INTEL in 2007. Those that are interested in a fuller bibliography that the references provide at the end of the book might like to look at Ian Clark's excellent website:

http://iangclark.net/metastability.html

David J. Kinniment
Newcastle, UK

List of Contributors

Although contributions to individual chapters have not been specified, four of my colleagues have provided substantial input to this book and their part in its preparation is gratefully acknowledged here.

Alexandre Bystrov	University of Newcastle upon Tyne
Marc Renaudin	INPG TIMA Laboratory, Grenoble
Gordon Russell	University of Newcastle upon Tyne
Alexandre Yakovlev	University of Newcastle upon Tyne

Acknowledgements

The authors would like to acknowledge the many contributions of Charles E. Dike, Intel Corp, arising from collaborative work between Intel and Newcastle University, and supported by UK EPSRC research grant EP/C007298/1. The help of Ran Ginosar and Mark Greenstreet in discussing the issues, and at Cambridge and Newcastle Universities, Robert Mullins, Fei Xia, Enzo D'Alessandro and Jun Zhou in commenting on several of the chapters is also gratefully acknowledged.

Some of the material in this book has been based on work published in the following papers and is reproduced by permission of the IEEE:

Synchronization Circuit Performance, by D.J. Kinniment, A. Bystrov, and A.V. Yakovlev which appeared in *IEEE Journal of Solid-State Circuits*, 37(2), 202–209 © 2002 IEEE.

Analysis of the oscillation problem in tri-flops, by O. Maevsky, D.J. Kinniment, A.Yakovlev, and A. Bystrov which appeared in *Proc. IS-CAS'02*, Scottsdale, Arizona, May 2002, IEEE, volume I, pp 381–384 © 2002 IEEE.

Priority Arbiters, by A Bystrov, D.J. Kinniment, and A. Yakovlev which appeared in *ASYNC'00*, pp 128–137. IEEE CS Press, April 2000 © 2002 IEEE.

Measuring Deep Metastability, by D.J. Kinniment, K Heron, and G Russell which appeared in *Proc. ASYNC'06*, Grenoble, France, March 2006, pp 2–11 © 2006 IEEE.

A Robust Synchronizer Circuit, by J Zhou, D.J. Kinniment, G Russell, and A Yakovlev which appeared in *Proc. ISVLSI'06*, March 2006, pp 442–443 © 2006 IEEE.

Multiple-Rail Phase-Encoding for NoC, by C. D'Alessandro, D. Shang, A. Bystrov, A. Yakovlev, and O. Maevsky which appeared in

Proc. ASYNC'06, Grenoble, France, March 2006, pp 107–116 © 2006 IEEE.

Demystifying Data-Driven and Pausible Clocking Schemes, by R Mullins and S Moore which appeared in *Proc. 13th Intl. Symp. on Advanced Research in Asynchronous Circuits and Systems (ASYNC)*, 2007 pp 175–185 © 2007 IEEE.

Efficient Self-Timed Interfaces for crossing Clock Domains by Chakraborty and M. Greenstreet, which appeared in *Proc. ASYNC2003*, Vancouver, 12–16 May 2003, pp 78–88 © 2003 IEEE.

1

Synchronization, Arbitration and Choice

1.1 INTRODUCTION

Digital systems have replaced their analog counterparts in computers, signal processing and much of communications hardware at least partly because they are much more reliable. They operate in a domain of absolutes where all the data moving from one place to another is quantized into defined high or low voltages, or binary coded combinations of highs and lows. The great advantage of this is that the data represented by the voltages can be preserved in the presence of small amounts of noise, or variations in component or power supplies, which can affect the voltages. A small variation in a high is still easily distinguished from a low, and a simple circuit, such as a gate or an inverter, can easily restore the correct output level.

In a synchronous system the signal states are also held by flip-flops when the clock ticks. Clocking digitizes time into discrete intervals of time, which are multiples of the clock cycle, with data computed during the first part of the cycle, and guaranteed correct on the next clock rising edge. Small variations in the time taken to evaluate a partial result have no effect on a correctly designed digital system, the data is always correct when it is next clocked. This discipline of two discrete voltage levels and time measured in multiples of the clock period allows most digital components to operate in a controlled environment, provided their outputs do not vary by more than a certain amount from the specified highs and lows, and their output times change only within

Synchronization and Arbitration in Digital Systems   D. Kinniment
© 2007 John Wiley & Sons, Ltd

a narrow band of times, a processor built on the components can be shown to operate correctly.

While this idea has worked well in electronic design for over 50 years the ideal world of a single system of discrete voltage levels and time intervals is not the real world. Within a processor binary data can be moved from one place to another under the control of a single clock, but to be useful the processor must be able to interact with inputs from the outside world, and to send its results back to the real world. Events in the real world are uncertain and unpredictable, the voltage levels from sensors are, at least initially, not the standard high or low voltage, they can occur at any time and can be any voltage level. We can try to hide this problem by using an analog-to-digital converter component which attempts to provide only standard outputs from analog inputs, but this can only ever be partly successful, since it involves a decision when an input is approximately halfway between two possible discrete output representations and one or the other output must be chosen. The mapping of input signal changes to discrete times also presents a problem. When a new input occurs that needs attention from a digital processor, the closed world of the processors discrete voltages and times must be interrupted so that the input can be dealt with. Again, there is a choice, which one out of an integral number of discrete clock cycles should be used to start the processing of that input? Since the input change can occur at any time, there will be occasions when the input is just before the next clock tick, and occasions when it is just after. If it is just before, it gets processed on the next tick, if it is after; it waits until the following tick to be noticed. The problem comes when it happens very close to the clock tick, because then there are two possibilities, this cycle, or the next cycle, and both of them are more or less equally desirable.

1.2   THE PROBLEM OF CHOICE

It looks as if it is easy to make the choices, and it is not a new problem. Whether you can make the choice in an acceptable time, and what it is that determines the choice it something that worried philosophers in the middle ages [1]. One of the first clear statements of the arguments was given by the Arab philosopher Abu Hamid Muhammad Ibn Muhammad al-Tusi al-Shafi'i al-Ghazali who wrote around 1100 AD:

> Suppose two similar dates in front of a man who has a strong desire for them, but who is unable to take them both. Surely he will take one of them through

a quality in him, the nature of which is to differentiate between two similar things.

He felt that it was obvious that a man or woman would be able to make the choice, and more specifically, to make the choice before starving to death from lack of food. But in order to decide you have to have some basis for choosing one course of action or another when they appear to be equal in value. His 'differentiating quality' was free will, which was thought to be inherent in mankind, but not in animals or machines.

The counter-argument is best given by Jehan Buridan, Rector of Paris University around 1340, in his example of the problem of choice which is usually known as the paradox of Buridan's Ass. In this example a dog is presented with two bowls, or an ass with two bales of hay. It has to choose one of them to eat or else starve to death. Animals were chosen deliberately here because they were not thought to be capable of free will, but even so, he discounts free will as the determining factor writing:

> Should two courses be judged equal, then the will cannot break the deadlock, all it can do is to suspend judgment until the circumstances change, and the right course of action is clear.

His insight was that the time required for a difficult decision would depend on the evidence available to make it, and if there is insufficient evidence it takes a long time. The implication is that there would be cases when even a man could starve to death because he could not decide.

1.3 CHOICE IN ELECTRONICS

As a philosophical discussion, the problem of choice without preference does not have much impact on real life, but it began to acquire real importance in the 1950s when the first digital computers were designed. For the first time many inputs could be processed in a very short space of time, and each input had first to be synchronized to the processor clock [2].

The solution seems obvious, if a digital signal represents the presence of an input, then it may appear at any time, but it can be synchronized to the clock with a simple flip-flop. This circuit is shown in Figure 1.1. When the system clock goes high, the output of the flip-flop will go high if the input was present, and not if it was absent so that the request can

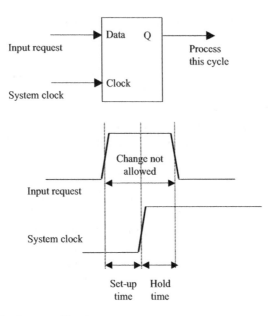

Figure 1.1   Flip-flop specification.

be processed in the next clock cycle. The difficulty with this solution is that the input request timing may violate the conditions for which the flip-flop specification is valid. A flip-flop data input must not change after a specified set-up time before the clock rising edge, or before a specified hold time after the clock edge. If it does, any guarantee that the output time will be within the specified times cannot hold.

In fact, as the input change gets closer to the clock edge, the flip-flop takes and longer to respond because the energy supplied by the overlap between data and clock inputs gets less and less, so it takes longer and longer to decide whether or not to interrupt the processor. Eventually, it reaches a level that is just sufficient to bring the Q output of the flip-flop to half way between a high, and a low value. At this point the flip-flop output is exactly balanced between high and low when the clock edge has finished, with no determining factor to push it one way or the other, so it can stay there for a relatively long time. This halfway state is known as a metastable state because it is not stable in the long term, it will eventually go on to be high or fall back to low, but in the metastable state the circuit has no drive towards either the high or low output values. The final outcome may be decided by internal noise, and though a metastable circuit will eventually settle in a well-defined state, the time taken is indeterminate, and could be very long.

If the input time cannot be constrained to change before the set-up time and after the hold time, then the flip-flop output time cannot be constrained either, so all output times relative to the clock also have a finite probability. This can have a knock on effect. If the output can change at any time between one clock edge and the next, it has not been truly synchronized to the clock, and might change part way through a clock cycle. This breaks the rules for a reliable digital system, with the result that systems may fail unpredictably because the processing of the interrupt may not be correct. The probability of a failure in the synchronization process can be very low for an individual interrupt, but because digital processors have to deal with very many inputs per second, the probability of catastrophic failure over a long period of time is not low, and must be at least quantified. It can only be eliminated by removing the need to synchronize, or reduced by allowing more time for the synchronizer to settle. Since there is always a small possibility of a very long settling time we must accept the possibility of failure in systems with a fixed clock period.

This fundamental problem was known to a few people in the early years of computing [2,3], but many, if not most engineers were not aware of the problems presented by synchronizers. In 1973 a special work-shop on synchronizer failures was held by Charles Molnar, Director of the Computer Systems Laboratory of Washington University, St Louis to publicize the problem and following this, more people began to accept that computers had been, and were being designed that were unreliable because the designers did not fully understand the problem of metastability. In an article in *Scientific American* that year a Digital Equipment Corporation engineer is quoted as saying 'ninety-nine out of a hundred engineers would probably deny the existence of the problem. They believe that conservative design is all you need; you simply allow enough time for a flip-flop to reach a decision. But as computers are designed to run faster and faster the safety factor gets squeezed out... the problem looks irresolvable.' Even as late as the 1980s a major glitch in the early warning missile radar system was attributed to a poor understanding by designers of the issue of synchronization.

1.4 ARBITRATION

A related problem occurs even if there is no clock. Without a clock, there is no need for synchronization between the input and the clock, but there is a need for arbitration between two competing inputs. Suppose two entirely seperate processors are competing for access to a memory, and only one can be allowed access at a time. The first to request access might get it, and when the memory access is complete, the second is allowed in. But if both make the request within a very short space of time, something must arbitrate between the requests to decide which gets the first access and which the second. As far as the hardware is concerned arbitration is much the same as synchronization, instead of deciding whether the request or the clock tick came first, we are deciding between two (or more) requests. It is only different in the way the arbiter output is treated. If there is no clock in the memory, it does not matter when the arbiter decides which input to accept, provided the data returned by the memory can be processed by the requester at any time. An arbiter can be designed which does not ever fail as a synchronizer might do, because it does not have to make the decision within a clock cycle. As the two requests get closer and closer in time, the arbitration takes longer and longer, and since the requests can occur within an infinitesmally small time, the response can take an infinitely long time. Instead of a failure rate there is a small probability of very long times, which may matter just as much in a real-time application as small probability of failure. Does it matter if a plane flies into a mountain because the navigation system fails, or it flies into the mountain because it cannot decide in time which way to go around?

1.5   CONTINUOUS AND DISCRETE QUANTITIES

Are we stuck with computer systems that may fail, or may not provide answers in the time we need them, or is there some circuit solution that will solve the problem? Unfortunately the basic problem has little to do with circuits. It is that real world inputs are continuous in both voltage and time. A continuous voltage input can have any value between the maximum and the minimum voltage level, say between 0 and 1 V, and there are an infinite number of possible values, 0.1, 0.2, 0.5 V, etc. between those two limits. On the other hand there are only two possible voltages for a binary coded signal, which might be 0 or 1 V. To decide which we must make a comparison between a fixed value, maybe 0.5 V, and the input. If the input is less than 0.5 V the output is 1, and if it is greater than 0.5 V the output is zero. It is possible to choose a voltage of 0.49 V as an input. This will be only 0.01 V different from the comparison value, so the comparison circuit has only a small voltage to work on. Any real physical circuit takes some time to produce an output of 1 V, and the smaller the input, the longer it takes. With a 0.4999 V input the circuit has an input of only 0.0001 V to work on, so it takes much longer to make the comparison, and in general there is no limit to the time that might be required for the comparison, because the input to the comparator circuit can be infinitesimally small.

Similarly, the processor interrupt can occur at an infinite number of instants between one clock edge and the next, and comparisons may have to be made on an infinitesimally small time interval. Again that could mean an infinite response time, and there will be a resulting probability of failure. To compute the failure rate you only need to decide the maximum time that the synchronizer has available to make its comparison, and then estimate how often the comparator input is small enough to cause the output to exceed that maximum.

1.6 TIMING

Synchronization of data passing between two systems is only necessary if the timing of the systems is different. If both systems work to the same clock, then they are synchronous and changes in the data from one system are always made at the same time in its clock cycle. The receiving system knows when the data is stable relative to its own clock and can sample it at that time so no synchronization is necessary. There are many possible clocking schemes for systems on a chip and Figure 1.2 shows three of them. In (a) a single clock is fed to all the processors, so that the whole system is essentially *synchronous*.

If the two clocks are not the same but linked, maybe through a phase locked loop, the relative timing might vary a little, but is essentially stable. Provided this phase variation is limited, it is still possible to transfer data at mutually compatible times without the need to use a synchronizer. The relationship is known as *mesochronous*. In (b), the individual processor clocks are all linked to the same source so that the timing is still synchronous or mesochronous. Often it is not possible to link the clocks of the sending and receiving system as they may be some distance apart. This is the situation in (c). Both clock frequencies may be nominally the same, but the phase difference can drift over a period of time in an unbounded manner. This relationship is called *plesiochronous* and there are times when both clocks are in phase and times where they are not. These points are to some extent predictable in advance, so synchronization can be achieved by predicting when conflicts might occur, and avoiding data transfers when the two clocks

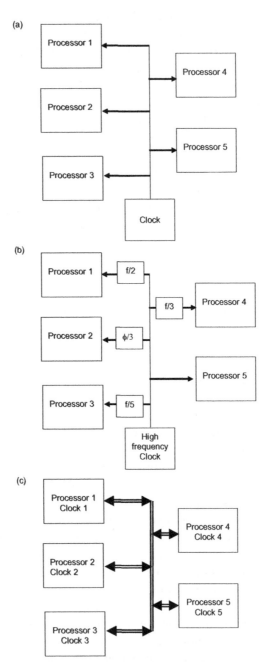

Figure 1.2 Clocking schemes: (a) single clock, multiple domains; (b) rational clock frequencies; (c) multiple clocks.

conflict. If the time frame of one system is completely unknown to the other, there is no way that a conflict can be avoided and it is essential that data transfers are synchronized every time. In this case the path is *heterochronous*. Heterochronous, plesiochronous and mesochronous are all examples of asynchronous timing.

1.7 BOOK STRUCTURE

The rest of this book has three main Parts. Part I is concerned with basic models of metastability in circuits, Part II with synchronization in systems, and Part III with arbitration.

In Part I, Chapters 2-5, mathematical models for metastability are derived which can be applied to simple synchronizer circuits with a view to explaining how the failure rates vary in different conditions. Having established the mathematical tools for comparing different circuits we present the key circuits used for synchronization and arbitration, the mutual exclusion element (MUTEX), the synchronizer Jamb latch, and the Q-flop. Modifications of these basic circuits to achieve higher performance and arbitration between more than two inputs are also described. Chapter 4 shows how thermal and other noise effects affect metastability, and the basic assumptions that must be satisfied in a system to allow the simple MTBF formulas to be used. Chapter 5 describes how metastability and synchronizer reliability can be measured. The metastability T_w and resolution time constant τ are defined and then methods for plotting typical histograms of failures against output time. From this we derive mean time between synchronizer failure in a system against synchronizer time, and input time against output time. Test methods suitable for on and off chip synchronizer measurement, both for custom ICs and field programmable devices, are discussed together with appropriate time measuring circuits. System reliability is dependent on metastable events occurring very late in the clock cycle, where the final state is determined by noise, or 'deep metastability'. Methods for measuring effects in this region including the effect of the back edge of the clock are explained.

In Part II, Chapters 6 and 7 discuss how synchronizers fit into a digital system. In Chapter 6, a system level view of synchronizers in presented. Where synchronization occurs in a multiply clocked system, how it can be avoided, and where it cannot be avoided, how to mitigate the effects of latency due to synchronizer resolution time are discussed.

High-throughput, low-latency, and speculative systems are described in Chapter 7, with the trade-offs between these two aspects of design. An alternative to multiple independent clocks is to use stoppable or pausible clocks. How these fit into a GALS system shown in Chapter 8. Chapter 9 concludes this Part.

Chapters 10–14 (Part III) present ideas about designing complex asynchronous arbiters using the building blocks studied in the previous chapters, such as synchronizers and mutual exclusion elements. These arbiters are built as speed-independent circuits, a class of circuits that guarantees their robust operation regardless of gate delays and any delay in handshake interconnects with the environment. A general definition for an arbiter is given and then a range of arbiters is covered, from simple two-way arbiters to tree-based and ring-based arbiters, finishing with a detailed examination of various types of priority arbiters, including static and dynamic priority schemes. The behaviour of arbiters is described throughout by Petri nets and their special interpretation signal transition graphs. The latter replace sets of traditional timing diagrams and provide a compact and formal specification of nontrivial dynamic behaviour of arbiters, potentially amenable to formal verification by existing Petri net analysis tools. The presented models and circuits enable solving complex arbitration problems encountered in a wide range of applications such as system-on-chip buses, network-on-chip and asynchronous distributed system in a more general sense.

Part I

2

Modelling Metastability

Asynchronous data originating from outside a clocked system has to be synchronized to the clock before it can be used. This usually involves a choice made on the basis of the arrival time of the data, which is a continuous variable, and the result of the decision, which clock edge to use, is discrete. In this situation it is possible to have very small input time differences between the data available signal and the clock. Such choices are difficult as the input energy to the synchronizer circuit falls towards zero when the time difference becomes small, and the response time of the circuit then approaches infinity. In fact there is always a point at which the synchronizer time response can become longer than the available time, and the synchronizer circuit fails because its output has not reached a stable state [4,5].

Asynchronous systems have a similar problem where the system must decide which of two or more asynchronous requests for a common processing resource is to be granted first. In this case the circuit that makes the decision is called an arbiter, and the input energy to the arbiter is also derived from the time differences between the inputs. Again, this time difference is a continuous variable and can also approach zero, with the result that the arbitration time can approach infinity [5,6]. In both cases the decision element, usually a flip-flop, has become metastable, and the resolution time of metastability in a flip-flop is important to the reliability of the system.

Synchronization and Arbitration in Digital Systems   D. Kinniment
© 2007 John Wiley & Sons, Ltd

2.1 THE SYNCHRONIZER

Most commonly metastability is seen in synchronization of data passing from one time domain into another. If the receiving domain is clocked, but the data input comes from a sending zone which may be clocked, but whose timing is unrelated to the receiver clock, we must first synchronize the data available signal to the receiving clock before any processing can be done. Typical synchronizer circuits use two edge-triggered flip-flops clocked by the receiving clock to achieve synchronization, as shown in Figure 2.1 with a time delay between the clocking of the first and second flip-flops.

The time delay is there to allow metastable levels in the first flip-flop to resolve before the second flip-flop is clocked, and therefore has a strong impact on the reliability of the synchronizer. The longer the flip-flop is given to resolve metastability the less likely it is that the synchronizer will fail. A flip-flop normally has two stable states, but the transition time between them is only guaranteed if the inputs conform to the set-up and hold times between data and clock. In the synchronizer the set-up and hold conditions will often be violated because the data available and clock timing are unrelated, so in a typical situation any value of time difference between them is possible. When the data input changes at a time close to the clock edge, the circuit outputs can be left in a metastable state, and the synchronizer designer needs to know how much time to allow in order to keep the chance of failure to an acceptable level

The edge-triggered flip-flops in the synchronizer are usually made from two level triggered latches in a master slave arrangement as shown in Figure 2.2. When the clock level is low, the master is transparent,

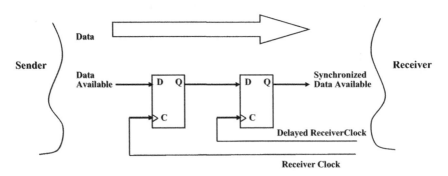

Figure 2.1 Synchronizer.

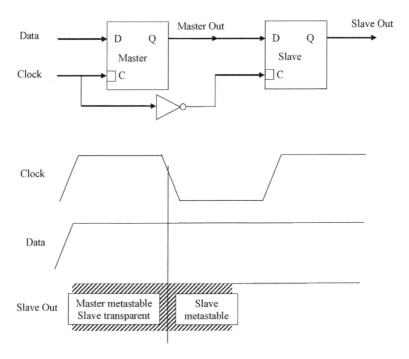

Figure 2.2   Edge-triggered flip-flop made from two latches.

that is, its data input is copied to the master output, but the slave is opaque and holds the previous value of the data input. When the clock goes high, the master becomes opaque, and the slave transparent. Because the clock and data times are unrelated in a synchronizer the set-up and hold times of the master latch are frequently violated. The master output may then go into a metastable state, and it may take some time to resolve to a well-defined digital level. While the clock is still high, any metastable level at the master output will also violate the conditions for the slave to act digitally and metastability can appear on the slave output. Since the problem is caused by master metastability the synchronization time will depend only on the metastability resolution time of the master. If the clock goes low before metastability is resolved in the master, the slave clock goes high at the same time and so the slave may now be metastable. From this point onwards any additional resolution time required depends on the characteristics of the slave.

To determine what these characteristics are, we need to examine what happens when the individual master and slave latch circuits become metastable.

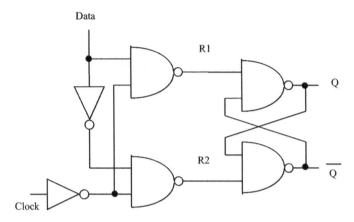

Figure 2.3   D latch.

A very simple design for a latch is shown in Figure 2.3. When the clock is high, both R1 and R2 are high, and the latch is opaque, that is, it is not affected by the data input and simply retains its previous state. When the clock goes low, R2 follows the data input, and R1 is the inverse of the data input. If the data input is high, R2 must be high, and Q is high. If the data input is low R1 is high, R2 low, and Q becomes high.

There are two ways that the circuit can become metastable, either the inputs are not digital levels, or the timing conditions are not observed. In the synchronizer circuit it the data can go high just before the clock goes high, so in this circuit there may be a runt pulse on R1, followed by both R1 and R2 going high. At this time the latch output Q may not yet have reached a high state, but the input from the D input is now inactive and only the cross-coupled feedback inputs can determine whether it ends up high or low.

Since R1 and R2 are now high, they take no further part in determining Q, so what happens then can be illustrated by considering a simple latch circuit made from two inverters.

This is shown in Figure 2.4(a). Here the input to the inverter on the left is the same as the output of the one on the right V_2, and the input of the right hand one is the output of the left hand one V_1. When the latch is metastable input and output voltages which are not just high or low can appear at the two points which are both inputs and outputs to the inverters. We can plot how the the output changes against the input for each inverter on different graphs, (b), or we can plot them both on the same graph by flipping the right-hand graph over and superimposing it on the line for the inverter on the left (c), so that only points, which lie on both input/output characteristics, are possible for both inverters in the long term.

One point has $V_1 = 0$ and $V_2 = 1$, and another has $V_1 = 1$ and $V_2 = 0$. These points are the two normal stable states. There is also a third point where $V_1 = V_2 = 1/2$, which is the metastable state.

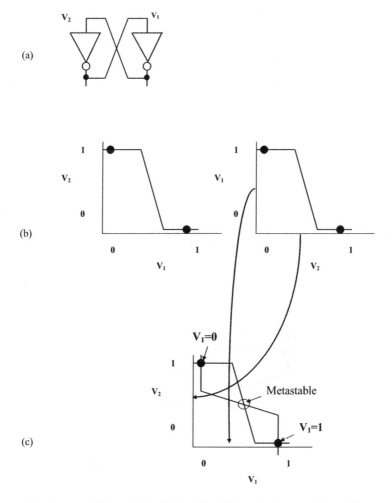

Figure 2.4 Metastability: (a) Latch bistable; (b) Static characteristics of inverters; (c) Superimposed characteristics.

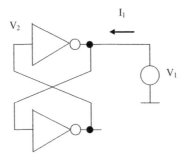

Figure 2.5   Output current against voltage.

To see this, suppose that the voltage V_1 is held by a voltage source, and then slowly increased from zero to V_{dd} as in Figure 2.5.

Suppose V_1 is held by the voltage source at a value of V_a volts. Because its input is at V_a, the output of the lower inverter in Figure 2.5 is at V_x. V_x at the input of the upper inverter would normally produce V_b at its output, but since it is held at a different voltage, V_a, a current is drawn from the voltage source, which is proportional to $V_a - V_b$. Now let's change the voltage V_1 slowly from zero to V_{dd}. The current first increases, as the difference between V_a and V_b increases, and then decreases to zero at the central point. Beyond that point, current has to be pulled out of the of the upper inverter to hold its output down, and this current only drops to zero at the second stable point where V_1 is high. The two superimposed characteristics are shown in Figure 2.6,

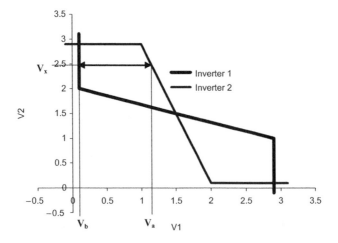

Figure 2.6   Superimposed characteristics.

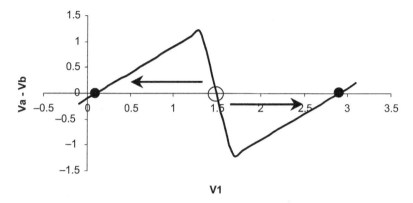

Figure 2.7 $V_a - V_b$ drive for different values of V_1.

and the resulting plot of $V_a - V_b$, which is a measure of the current, against V_1 is shown in Figure 2.7.

In a real circuit there is always some stray capacitance associated with each node, so with a capacitor connected to the output rather than a voltage source, we can see what will happen when the node is metastable. In Figure 2.7 the capacitor is discharged when V_1 is on the left-hand side of the central point, pulling V_1 lower, and charged if it is on the right-hand side, pulling it higher. The central point is not stable. This kind of instability is analogous to a ball placed on top of a hill (Figure 2.8) if we make the slope of the hill the same as Figure 2.7, i.e. positive on the left-hand side and negative on the right-hand side, there are two stable points where the slope is zero, on either side of the central, unstable point where the slope is also zero. When the ball is let go it will eventually roll down to one side or the other, but with careful placement it can stay in the metastable state at the top for some time.

The analogy of Figure 2.8 shows that the output moves rapidly to either the $V_1 = 1$ and $V_2 = 0$ state or the $V_1 = 0$ and $V_2 = 1$ state, the differences between the actual operating trajectory and the static characteristics being accounted for by the output capacitance which absorbs the difference current. The smaller the capacitance, the faster it goes. The central point is usually called the metastable point, and a latch is said to be metastable if it is in balance between a 1 and a 0, with the drive towards the two stable points, called attractors, close to equal. The difficulty with metastability is that with a net drive of zero, the capacitances are neither charged, nor discharged, and the circuit can remain at the metastable point for a long time, much longer than its normal response time with digital inputs.

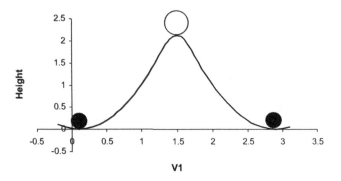

Figure 2.8 The ball on the hill.

If the input clock/data overlap is very close to a critical point, metastability can be reached from either stable state as shown in Figure 2.9. This particular photograph was taken by selecting all the metastable events in a level triggered latch, which lasted longer than 10 ns (two divisions) [6]. Several traces are superimposed, with outputs starting from both high and low values, then reaching a metastable value about halfway between high and low, and finally going to a stable low level. It can be seen that the traces become fainter to the right, showing that there are some with long metastability times, but fewer than the short metastability time responses.

If only one flip-flop is used for synchronization, the master latch in the flip-flop will often be left in a metastable state and a long time may elapse before its output escapes to a stable high or low. A half level input, or a change in level as the clock changes in the slave may then

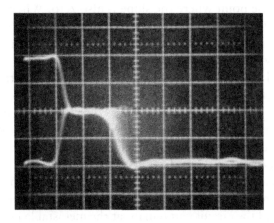

Figure 2.9 Output from a metastable latch.

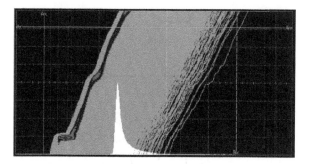

Figure 2.10 Synchronizer output times.

result in a long time delay in the output data available signal which is first read by following circuits as a low level, and then later as high.

Figure 2.10 shows the output of a CMOS flip-flop used in a synchronizer situation. Many outputs have been captured on a digital oscilloscope at 1 ns/division, and though the circuit is specified to have a response time of around 3 ns some of the high-going outputs are 5 ns or more after the clock time. As time increases from left to right, the density of the traces reduces, because the probability of the output changing reduces. A histogram of the number of outputs changing at a particular time is also shown in this figure (the white area, in which the height at a particular time is proportional to the number of outputs crossing the A_x line at that time). If the output changes late, there may be synchronizer failures, but the longer the delay allowed, the less likely is the synchronizer to fail by indecision. We can explain metastability qualitatively this way, but to produce reliable designs a more accurate quantified model is needed.

2.2   LATCH MODEL

In the metastable state both inverters making up the flip-flop are operating in a linear mode. If they were not, the circuit would never escape from metastability, and to compute its escape time it is necessary to analyse the analog response of the basic circuits used. Most models of flip-flop circuits operating as synchronizers assume that the cross-coupled gate circuits making up the flip-flop are in a small signal mode, never far from the metastable state. Here to make the analysis simpler by eliminating constants, we will assume that the metastable state of the gates is at 0 V, rather than $V_{dd}/2$. This means that a digital high output is $+ V_{dd}/2$, and a low output is $- V_{dd}/2$.

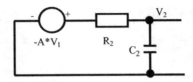

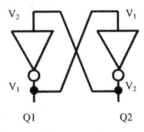

Figure 2.11 Small signal models of gate and flip-flop. Reproduced from Figure 13, "Synchronization Circuit Performance" by D.J.Kinniment, A Bystrov, A.V.Yakovlev, published in IEEE Journal of Solid-State Circuits, 37(2), pp. 202–209 © 2002 IEEE.

The gates can now be modelled as two linear amplifiers [6–8]. Each gate is represented by an amplifier of gain $-A$ and time constant CR, as shown in Figure 2.11. Differing time constants due to different loading conditions can also be taken into account [8].

The small signal model for each gate has a gain $-A$ and its output time constant is determined by C/G, where G is the gate output conductance and is equal to $1/R$. In a synchronizer, both the data and clock timing may change within a very short time, but no further changes will occur for a full clock period, so we can also assume that the input is monotonic, and the response is unaffected by input changes.

For each inverter we can write:

$$-C_2\frac{dV_2}{dt} = G_2V_2 + AG_2V_1$$

$$-C_1\frac{dV_1}{dt} = G_1V_1 + AG_1V_2$$

(2.1)

The two time constants can be simplified to:

$$\tau_1 = \frac{C_1}{AG_1}, \; \tau_2 = \frac{C_2}{AG_2}$$

(2.2)

Eliminating V_1, this leads to

$$0 = \tau_1.\tau_2.\frac{d^2V_1}{dt^2} + \frac{(\tau_1+\tau_2)}{A}\frac{dV_1}{dt} + \left(\frac{1}{A^2}-1\right)V_1 \qquad (2.3)$$

This is a second-order differential equation, and has a solution of the form:

$$V_1 = K_a\, e^{\frac{-t}{\tau_a}} + K_b\, e^{\frac{t}{\tau_b}} \qquad (2.4)$$

In most practical cases, the inverters have a high gain, and are identical, so $A \gg 1$, $\tau_a = \tau_b = \sqrt{\tau_1\tau_2}$.

Both K_a and K_b are determined by the initial conditions, τ_a and τ_b by τ_1, τ_2, and A. Typical values of τ_1, τ_2, and A for a, 0.18 μ, process, are 35 ps for τ_1 and τ_2 (depending on the loading of the inverters) and 20 for A. Usually the values of τ_1 and τ_2 are similar to those of an inverter with a fan-out of 4, since both times are determined by the capacitances, conductances, and gain of the transistors in the circuits.

This model is valid within the linear region of about 50 mV either side of the metastable point. Outside this region the gain falls to less than 1 at a few hundred mV, but the output resistance of the inverter and the load capacitance also drop significantly, R by a factor of more than 10, and C by a factor of about 2. Thus, even well away from the metastable point the values of τ_1 and τ_2 still have values similar to those at the metastable point.

2.3   FAILURE RATES

Using the model of metastability above, it is possible to estimate the input conditions necessary to produce a metastable response of a given time, and hence how often this time will be exceeded in a synchronizer. If we assume that we are only interested in metastable events that take a much longer time than the normal flip-flop response time, we can neglect the first term in Equation (2.4) for the output voltage and use:

$$V_1 = K_b\, e^{\frac{t}{\tau}} \qquad (2.5)$$

The initial condition K_b depends on the input time overlap between clock and data. If the data input changes to a high much earlier than the clock, K_b will be positive, so the output voltage will reach the digital high value of $+ V_{dd}/2$ quickly. If it changes from low to high much later than the clock K_b will be negative, so the output voltage will reach a low value quickly. In between, the value of K_b will vary according to the relative clock data timing, and at some critical point $K_b = 0$, so the output voltage is stuck at zero. We will call the clock data timing that gives $K_b = 0$, the balance point, where in the previous analogy, the ball is exactly balanced on top of the hill and we will measure input times relative to this point using the symbol Δt_{in}. The value of K_b is given by:

$$K_b = \theta \, \Delta t_{in} \qquad (2.6)$$

where θ is a circuit constant which determines the rate at which the clock data overlap converts into a voltage difference between the two nodes. If the cross coupled gates in the flip-flop are 2 input NAND gates either V_{dd}, or $0\,V$, and are connected via a conductance G_s to the output capacitor C for a time Δt_{in} as shown in Figure 2.12, the output node will charge or discharge, moving its voltage away from the central point $V_{dd}/2$. G_s is usually less than AG, and $\tau_s = C/G_s$ is similar to the rise or fall time of that gate, so $\theta \approx V_{dd}/2\tau_s$.

Metastability is over when the latch output has reached normal levels, either high or low. The time taken to reach $|V_e|$, the exit voltage where the trajectory leaves the linear region and can be regarded as a stable high, or low voltage, can be calculated from Equation (2.5) by putting $V_1 = V_e$. Since $K_b = \theta \Delta t_{in}$, this means that the exit time is given by:

$$t = \tau \, \ln\left[\frac{V_e}{\theta \, \Delta t_{in}} \right] \qquad (2.7)$$

If the timing relationship between data and clock changes is unknown, all values of Δt_{in} are equally probable. In these circumstances It is usual to assume that the probability of any input smaller than given overlap, Δt_{in}, is proportional to the size of the overlap, very small values being

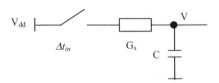

V_{dd} Δt_{in} G_s C V

Figure 2.12   Input time constant.

therefore much less common than large ones. This is usually true if
the timing of both send and receive clocks are independent oscillators.
When the sending data rate is, on average f_d and the receiving clock
frequency f_c, each data item sent could occur at any time between 0
and $1/f_c$ of the next receiving clock edge, so the probability of that data
available signal having an overlap of Δt_{in} or less, is $f_c \Delta t_{in}$.

Within some long time T there will be $f_d T$ data items transmitted, so
the probability of at least one of the overlaps being less than Δt_{in} is $f_d f_c$
$T \Delta t_{in}$.

If the synchronizer fails when the input overlap is shorter than Δt_{in} by
giving an output response longer than the resolution time allowed, the
mean time between failures will be:

$$MTBF = \frac{1}{f_d f_c \Delta t_{in}} \tag{2.8}$$

Failure occurs when

$$\Delta t_{in} = \frac{V_e}{\theta} e^{-\frac{t}{\tau}}$$

so we can get the mean time between failures of the synchronizer as:

$$MTBF = \frac{e^{\frac{t}{\tau}} \theta}{f_d f_c V_e} \tag{2.9}$$

This is more usually written as

$$MTBF = \frac{e^{\frac{t}{\tau}}}{f_d f_c T_w} \tag{2.10}$$

where $T_w = V_e / \theta$, and T_w is known as the metastability window.

MTBF depends on f_c and f_d, which are system parameters rather
than circuit parameters, so we can use Equations (2.8) and (2.10) to
get a characterization of a synchronizer which is dependent only on
the circuit parameters τ and T_w:

$$\Delta t_{in} = T_w e^{-\frac{t}{\tau}} \tag{2.11}$$

The metastability window is the input time interval for which output
times are longer than normal. If $0 < \Delta t_{in} < T_w$, then the output response
time to V_e, t, is greater than zero, and the circuit is metastable. Input

time differences outside this range either cause a normal response or no change. The value of T_w is determined by the input time constant θ and the point at which the flip-flop exits from metastability V_e; thus, if

$$V_e = \frac{V_{dd}}{2}$$

T_w is the same order of magnitude as the rise time of the gate. The value of τ is mainly determined by the feedback loop time constants, and since both T_w and τ are determined by channel conductances and gate capacitances, they are likely to be similar. This analysis applies to most simple latches, but may not hold for more complex devices, made from gates with more than one time constant in the feedback loop, or with long interconnections. It is very important for any flip-flop used for synchronization that it is characterized as a single cell with a fixed layout, and not assembled from individual gates or, for example, FPGA cells, because the feedback interconnection may have additional time constants, and the differential equation that describes the small signal beviour will be correspondingly complex. An example of multiple time constants is shown in Figure 2.13, where a latch has been assembled out of two FPGA cells. In this figure the latch output change is being used to trigger the oscilloscope, and the figure shows a histogram of clock events. The time scale is 0.5 ns/division. Because of the additional delay in the feedback loop the resolution time constant is nearly 1ns, and the histogram of clock to output times rises and falls as a result of oscillation in the output trajectory. For comparison the histogram from a standard

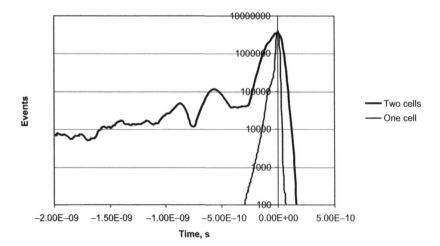

Figure 2.13   Multiple FPGA cell latch.

latch on the same FPGA is also shown. Here, the resolution time constant is much faster at around 40 ps and there is no sign of oscillation.

If there is enough delay in the feedback loop, as is the case with latches constructed from more than one cell, the amplitude of the oscillation produced can be sufficient to take the gates outside the linear region, and the oscillation can look like a square wave, which persists for much longer than the times shown even in Figure 2.13.

If the time allowed for metastability is short, we must also take into account the first term in the response equation. For many metastable events, the initial conditions are such that K_b, which is the voltage difference between output nodes at the start, is less than 10 mV, and while K_a, the common offset at the start, may be as much as 0.5 V. In these circumstances the K_a term is important, but only in determining the response times for metstable events resolving early. This is because simple circuits using gates with only one time constant, always reach the linear region quickly. τ_a is small outside the linear region and so the K_a term becomes small very quickly. The metastable events that cause synchronizer failures last a long time, and will have trajectories that spend most of their time in the linear region and so we can usually still use the simplified second-order model to predict the time it takes flip-flop circuits to escape from metastability.

The difference between early and late trajectories, and how they are affected by the K_a term can be seen in Figure 2.14.

This figure shows the response of the two outputs in a latch against time as predicted by the small signal model when both outputs start from a point higher than the metastable level and from a point lower. There is a common mode initial offset of both outputs given by $K_a = +150$ mV

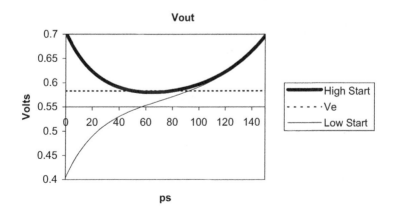

Figure 2.14 Model response. Reproduced from Figure 2, "Synchronization Circuit Performance" by D.J.Kinniment, A Bystrov, A.V.Yakovlev, published in IEEE Journal of Solid-State Circuits, 37(2), pp. 202–209 © 2002 IEEE.

and $-150\,\text{mV}$ from the metastable level of $0.55\,\text{V}$ for the high start and low start curves respectively. For both these trajectories $K_b = 4\,\text{mV}$, representing the initial difference between the two outputs. We use $\tau_a = 25\,\text{ps}$ and $\tau_b = 42.5\,\text{ps}$ to represent a typical situation where A is about 5–10.

It is common for the exit from metastability to be detected by an inverter with a slightly different threshold from the metastability level of the flip-flop. Thus when V_{out} exceeds that level the inverter output changes from a high to a low. The metastability level of the flip-flop here is $0.55\,\text{V}$, and the threshold level is $0.58\,\text{V}$ (the dotted line). V_{out} exceeds $0.58\,\text{V}$ at $80\,\text{ps}$ for the high start of $0.7\,\text{V}$ and at $93\,\text{ps}$ for the low start of $0.4\,\text{V}$. Note that the time difference depends upon the threshold level, and that if the high start trajectory never goes below $0.58\,\text{V}$, that is if $K_b > 4.7\,\text{mV}$, metastability is not detectable for output time delays between 0 and $65\,\text{ps}$ because the detecting inverter output remains low all the time. From these curves it can be seen that the metastability resolution time depends on the value of K_a during the first $100\,\text{ps}$, but not beyond that.

2.3.1 Event Histograms and MTBF

Measurements of flip-flop performance in metastability have been done in at least two different ways.

The most common measure is to count the number of times in a given period that metastability lasts longer than a particular time t. This is a measure of MTBF when the resolution time is t (Equation 2.10). If T is the time period over which the events are counted, MTBF = T/(total events) (Equation 2.12). A plot of the total events lasting longer than a particular resolution time for a synchronizer flip-flop is shown in Figure 2.15 where it is compared with the second measure.

$$\text{Total events} = \frac{T}{MTBF} = f_d\, f_c\, T\, T_w\, e^{\frac{-t}{\tau}} \qquad (2.12)$$

The alternative is to count the number of times the output escapes from metastability within δt of t (events). Again T is the period over which events are measured, and the number of events that resolve within δt is

$$-\frac{\text{d}}{\text{d}t}\left[f_d\, f_c\, T_w\, e^{\frac{-t}{\tau}} \right] \delta t$$

or

$$\frac{\delta t}{\tau}\left[f_d\, f_c\, T\, T_w\, e^{\frac{-t}{\tau}}\right] \tag{2.13}$$

These events are also shown in Figure 2.15. Because the events scale is logarithmic, and many more events appear if the resolution time is only slightly reduced, the total events curve is only slightly above the events curve in most places. The main differences between the two sets of results are that small variations in the slope of the curve tend to be slightly exaggerated in the events curve, so that the change in slope shown in Figure 2.15 at around 6 ns appears to be sharper, and the two curves diverge for output times below the normal propagation time of the flip-flop.

The same data can be presented in a more useful form (Figure 2.16) which shows the MTBF against resolution (synchronization) time as well as the input time required to give a particular resolution time. Using Equation (2.12) total events can be divided by the time T, during which the events were collected to give $1/MTBF$, and (2.11) enables the value of Δt_{in} to be calculated. Graphs like this are also used to find the values of the circuit parameters τ, and T_w. If at any point we draw a tangent to the curve (shown by the dashed line in Figure 2.16) the slope of the tangent is $-1/\tau$, so τ can be calculated. Here it is about 125 ps, and the intercept on the Δt_{in} axis is T_w. T_w can be measured at the normal propagation time, or at $t = 0$, in which case the value of $T_{w0} = 10^7$ s here, but if it is measured from the normal propagation time of the flip-flop, about 3.9 ns, $T_w = 3 \times 10^{-7}$.

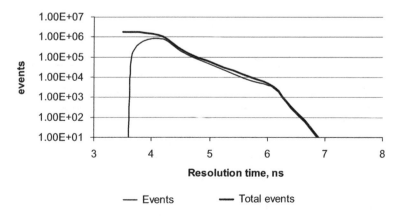

Figure 2.15 Events against resolution time.

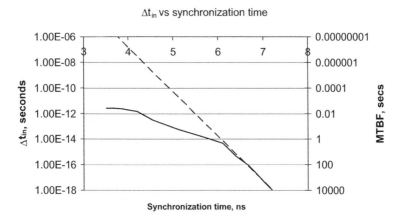

Figure 2.16 *MTBF*, τ, and *T*$_w$.

The methods used to measure event histograms from typical circuits are described in more detail in Section 4.2.

Figure 2.17 shows how an initial offset affects the measurement of metastability. In Figure 2.17, $K_a = +150\,\text{mV}$ for the high start curve, and $-150\,\text{mV}$ for the low start, both with $\tau_a = \tau_b = 33\,\text{ps}$. This is a histogram of the probability of an output trajectory escaping metastability within 3.3 ps of a particular output time. In order to calculate the model responses in Figure 2.17 a uniform distribution of voltage offsets K_b has been derived from a distribution of input time differences by using the relationship between K_b and Δt_{in} given in Equation (2.6), where $\theta = 10\,\text{mV/ps}$ in this case.

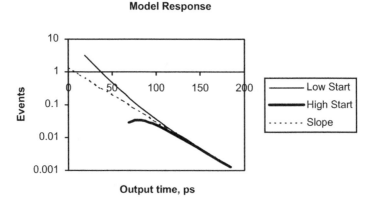

Figure 2.17 Events per unit time as a function of metastability time. Reproduced from Figure 3, "Synchronization Circuit Performance" by D.J.Kinniment, A Bystrov, A.V.Yakovlev, published in IEEE Journal of Solid-State Circuits, 37(2), pp. 202–209 © 2002 IEEE.

In a real experiment the number of events at a particular time is given by Equation (2.13). In Figure 2.17, $\tau = 33$ ps, $T_w = V_e/\theta = 33$ mV/10 mV/ ps $= 3.3$ ps, and $\delta t = 3.3$ ps, so this figure shows the output event histogram against output time for the low start and the high start situations corresponding to the two values of K_a.

We use $T = 3.3$ ms, f_c and $f_d = 30$ MHz, to get a number of events (in this case a probability of a number of events if the number is less than one) of around 1 when $t = 0$, though in practice a time T of seconds would give many more events. The two curves deviate from the expected straight-line relationship, the high start curve recording no events before 65 ps because trajectories with large enough initial conditions do not intersect the output threshold of 0.58 V, and then events are recorded earlier than expected.

In the low start curve, events are delayed rather than accelerated by the effects of the $K_a e^{-t/\tau_a}$ term in Equation (2.4), so the event histograms can look quite different at low values of t, though they converge for high values. This is important if we are going to use the slope of early events to predict $MTBF$ values for much longer synchronization times, and we must use the full expression including K_a as well as K_b for calculating the mean time between synchronizer failures. Putting V_s for the initial common mode offset voltage V_e for the final voltage where the trajectory leaves the linear region, as before, we have:

$$V_e = V_s \, e^{\frac{-t}{\tau_a}} + \theta \, \Delta t_m \, e^{\frac{t}{\tau_b}} \tag{2.14}$$

Hence, by the same argument that we used earlier, we can show:

$$MTBF = \frac{e^{t/\tau_b}}{\left[\dfrac{V_e - V_s e^{\frac{-t}{\tau_a}}}{\theta}\right] f_c \, f_d} \tag{2.15}$$

Now

$$T_w = \frac{V_e - V_s e^{\frac{-t}{\tau}}}{\theta}$$

and is no longer constant. In many circuits the initial offset is not negligible, and so the initial value of T_w can be different from its final value; either higher, or lower, depending on the polarity of the difference between V_e and V_s, and it is dangerous to use early values of slope to calculate τ

2.4   LATCHES AND FLIP-FLOPS

If a synchronizer is made from two latches, its reliability depends straightforwardly on the time allowed for the first latch to resolve between the rise time of the first and the second latch clocks C1 and C2 as in Figure 2.18. The MTBF calculation given by Equation (2.10) applies. The throughput of the two latch arrangement is limited, as the input data can only be sampled when C1 goes high, and C1 cannot be taken low until C2 goes high. The resolution time available for the latch1 output is then only half the cycle time.

To increase the throughput, synchronizers are normally made from two edge-triggered flip-flops as shown in Figure 2.1. This allows a resolution time approximately equal to the cycle time, but calculation of the reliability is more complex. The reliability now depends on resolution of metastability in the master–slave combination, and the important factor is the probability of resolution before the end of the second half of the clock cycle, because that is when the synchronized output appears. In Figure 2.1 both FF1 and FF2 are made from two latches, and the output of FF1 can become metastable either as a result of metastability in the master of FF1 being copied through the slave in the first half of the clock cycle, or metastability in the slave being induced by set-up and hold

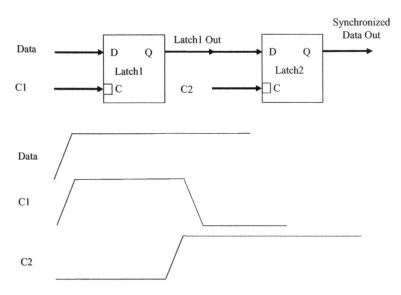

Figure 2.18   Latch-based synchronizer.

condition being violated for the slave of FF1 when the clock goes low at the beginning the second half of the clock cycle. Slave metastability is usually caused by metastable events at the output of the master latch in FF1 that resolve very close to the falling edge of the clock. The number of such events within a time δt of the clock half-period $t/2$ is given by Equation (2.13), and is:

$$\frac{\delta t}{\tau}\left[f_d\, f_c\, T\, T_w\, e^{\frac{-t}{2\tau}}\right] \tag{2.16}$$

Any metastability must be resolved by the slave latch of FF1 in the second half of the cycle, within a time $t/2$, so according to Equation (2.11) only events within a time Δt_{in} of the balance point of the latch will cause failures where:

$$\Delta t_{in} = T_w e^{-\frac{t}{2\tau}} \tag{2.17}$$

Putting $\Delta t_{in} = \delta t$ in Equation (2.16) we get the number of events that cause the synchronizer fail in a long period of time T, which is:

$$\frac{T_w e^{-\frac{t}{2\tau}}}{\tau}\left[f_d\, f_c\, T\, T_w\, e^{\frac{-t}{2\tau}}\right] = \frac{T_w}{\tau}f_d\, f_c\, T\, T_w\, e^{\frac{-t}{\tau}} \tag{2.18}$$

From this we can compute the $MTBF$ of the master slave combination:

$$MTBF = \frac{e^{\frac{t}{\tau}}}{f_d\, f_c\, T_w\, T_w}\tau \tag{2.19}$$

The form of Equation (2.19) is exactly the same as Equation (2.10), but the value of T_w for the edge triggered flip-flop combination is T_w^2/τ, which is slightly different from the value of T_w for its component latches.

As well as having a different T_w the flip-flop has a different behaviour when the input is outside the metastability window. For inputs much earlier than the clock, a latch is transparent, so that the output follows the input with a short delay and the output time can go negative. On the other hand the output of an edge-triggered flip-flop does not change until after the clock edge, so no output events are seen with times earlier than the clock edge time.

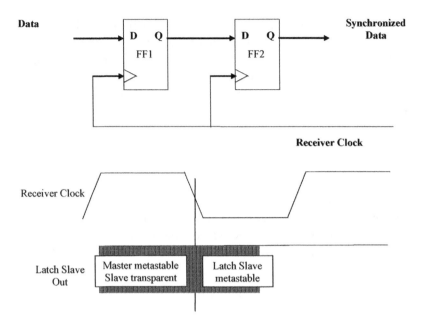

Figure 2.19   Edge-triggered flop-flop-based synchronizer.

This analysis assumes that both latches are identical, and does not take into account factors like the propagation delay of signals through a transparent latch which will affect the proportion of the clock cycle available for metastability resolution.

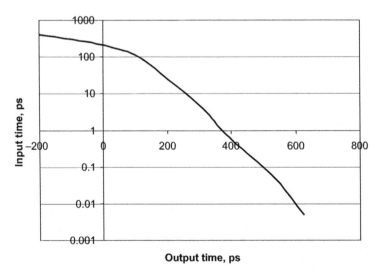

Figure 2.20   Latch input time vs output time showing negative values for output time.

2.5   CLOCK BACK EDGE

Because of the internal construction of an edge-triggered flip-flop, there is often a significant increase in the failure rate in the second half of the clock cycle due to the delay in the slave latch of the flip-flop. We can now extend the theory presented so far to show what happens when the back edge of the clock occurs.

Consider two similar master and slave level triggered latch circuits cascaded to form a single edge-triggered circuit as shown in Figure 2.21.

Here the master and slave can both be reset so that both Out1 and Out2 are low. When the clock is low, the master is transparent and any change in In1 is copied through to Out1 with a delay T_d determined mainly by internal large signal gate delays. In normal operation In1 does not change within the set-up time before the clock rising edge, or inside the hold time after the rising edge, thus Out1 is steady when the master latch goes opaque and input changes no longer have any effect. At around the same time as the clock rising edge, the slave clock falls and so the slave goes transparent. Now the Out1 level is transferred to Out2 with a delay T_d.

If the circuit is used as a synchronizer the input can go high at any time and so a rising edge on the input In1 which occurs just before the clock rising edge may cause the master to go into metastability. If the metastability is resolved well before the falling edge of the clock the change in Out1 is copied through the transparent slave to Out2 with the normal delay T_d and if it is resolved after the falling edge the timing of Out2 is unaffected. Because the change in Out1 and consequently In2 may happen very close to the falling clock edge of the master, an input change can fall in the metastability window of the slave. Consequently there is a low, but finite probability that metastability in the master can produce metastability in the slave.

The two latch circuits are normally similar, so that the metastability time constant τ, in both is about the same and it is often assumed that

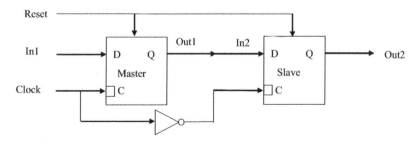

Figure 2.21   Edge-triggered synchronizer.

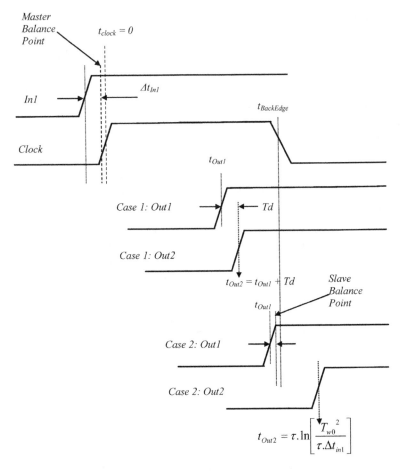

Figure 2.22   Waveforms of master and slave in metastability.

the total resolution time of a metastable output, t_{Out2} follows the natural logarithmic form described in Equation (2.7).

$$t_{Out2} = \tau \ln\left[\frac{T_{w0}}{\Delta t_{In1}}\right] \qquad (2.20)$$

In this equation Δt_{in1} is the input time relative to the balance point. In fact, this relationship does not quite hold for an edge-triggered flip-flop, as the following analysis shows. In Figure 2.22 the waveforms of the master and slave are shown when both are close to metastability. In this figure the rising edge of the clock occurs at a time $t_{clock} = 0$ and all times are measured with reference to this time. If the slave output occurs at a time, t_{Out1} late in the first half-cycle of the clock, its timing depends on the input time Δt_{in1}, before the balance point. The closer Δt_{in1} is to the balance point of the master, the later is t_{Out1}: We can plot Δt_{in1} against t_{Out1} for the latch on its own by observing that

$$t_{Out1} = \tau \ln \left[\frac{T_{w0}}{\Delta t_{In1}} \right] \tag{2.21}$$

From this it is possible to find the output time of the slave t_{out2}. There are two cases:

Case 1: t_{Out1} happens well before the back edge of the clock, so that the slave is transparent when its input changes. In this case the constant delay of the slave T_d is simply added to t_{Out1}, so t_{Out2} is given by:

$$t_{Out2} = T_d + \tau \ln \left[\frac{T_{w0}}{\Delta t_{in1}} \right] \tag{2.22}$$

Case 2: the slave may go metastable. In this case t_{Out1} must be very close to the absolute slave balance point time. We have already seen Equation (2.1) that in the second half of the cycle,

$$MTBF = \frac{e^{\frac{t}{\tau}}}{f_d f_c T_w T_w} \tau$$

so using the relationship between $MTBF$ and Δt_{in} in Equation (2.8) we can write

$$t_{Out2} = \tau \ln \left[\frac{T_{w0}^2}{\tau \Delta t_{in1}} \right] \tag{2.23}$$

Here, the balance point has changed. In case 1, the balance point was the input that gave equal probability of a high and low outputs in the master, but in case 2, it is the point that gives equal probability of a high and low outputs in the slave. The balance point for the master–slave combination is therefore the input time that gives a slave input exactly at the balance point of the slave.

$$\Delta t_{SBP} = T_{w0} \, e^{-\frac{t_{SBP}}{\tau}} \tag{2.24}$$

When compared with a latch, the balance point for the combination is shifted by this amount. Any plot of the input times against output

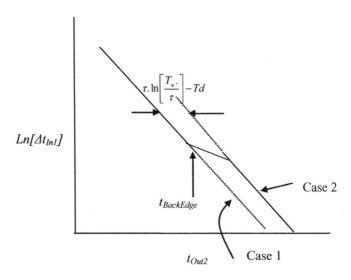

Figure 2.23 Case 1: slave transparent, case 2: slave metastable.

times for a flip-flop made from two latches must therefore be measured from this new point. The change makes little difference for values of for values of $t_{Out1} \ll t_{BackEdge}$, since

$$\Delta t_{In1} >> T_{w0}\, e^{-\frac{t_{BackEdge}}{\tau}}$$

but it becomes important when $t_{Out1} \approx t_{BackEdge}$.

Figure 2.23 Illustrates the input time against output time that may be expected from a master–slave flip-flop. Case 1, on the left shows the relationship of Equation (2.22) with a slope of τ, normally seen between Δt_{in1} and t_{Out2} when $t_{Out2} < t_{BackEdge}$. On the right, the slope of the curve given by Equation (2.23) when $t_{Out2} > t_{BackEdge}$ remains the same, but there is a displacement which is given by the difference between the two cases:

$$\tau \ln\left[\frac{T_{w0}\, T_{w0}}{\Delta t_{In1}\, \tau}\right] - Td - \tau \ln\left[\frac{T_{w0}}{\Delta t_{In1}}\right]$$

or simplifying,

$$\tau \ln\left[\frac{T_{w0}}{\tau}\right] - Td \qquad\qquad (2.25)$$

Here it can be seen that the offset depends on the circuit constants τ, T_{w0} and T_d.

3

Circuits

In the design of a synchronizer the values of T_w and τ depend on the circuit configuration. Both affect the resolution time, but τ is more important for synchronization because the synchronization time needed is proportional to the resolution time constant τ. The effect of increasing T_w by a factor A is simply to add to the synchronization time an amount equal to $\tau \ln(A)$.

T_w is mainly determined by the input characteristics of a latch circuit and τ is the time constant of the feedback loop. To some degree these two can be traded, a low-power input drive can reduce the loading on the feedback inverters thus reducing τ, but usually at the expense of T_w. Increasing power can often reduce both T_w and τ because the parasitic capacitances become a lower proportion of all capacitance, but only up to a point where parasitic capacitances become negligible and the value of C/G reaches a minimum.

3.1 LATCHES AND METASTABILITY FILTERS

Figure 3.1 shows a simple latch made up of four NAND gates and an inverter. When the clock goes low, both R1 and R2 go high and the latch becomes opaque. Without the two inverters on the outputs a metastable level of $V_{dd}/2$ could cause any following digital circuits to malfunction when the cicuit becomes metastable. The inverters on the outputs of Figure 3.1 prevent the half-level appearing at the output because they have a lower than normal threshold level. If the latch is in a metastable state, both inverter outputs are low because the output level when the

Synchronization and Arbitration in Digital Systems   D. Kinniment
© 2007 John Wiley & Sons, Ltd

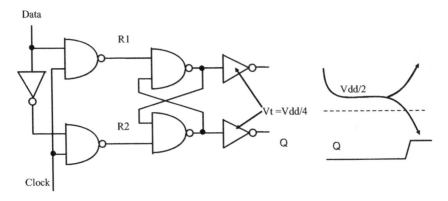

Figure 3.1 Latch with low-threshold inverters.

circuit is metastable is higher than the output inverter threshold. Only when one latch output moves lower than the inverter threshold can the corresponding inverter output go high. The *W/L* ratio of the transistors in the inverters is crucial here to make low-threshold inverters. Wider *n*-type transistors and narrow *p*-type will give a lower than normal threshold, and wider *p*-types than *n*-types give a higher threshold.

An alternative metastability filter arrangement [9] is shown in Figure 3.2, where a high output only appears when there is sufficient difference between the two latch outputs. The advantage of this is that it will filter out metastable outputs where both output voltages have the same value irrespective of whether the outpt voltage is, high, low, or

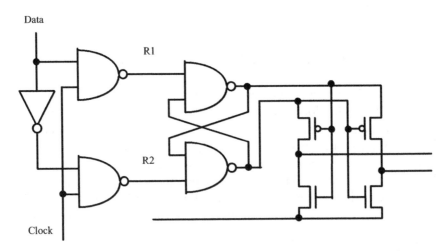

Figure 3.2 Latch with metastability filter.

half-level. Thus it can also be used to remove the effects of metastability in circuits like Figure 2.13 where both outputs go up and down in phase. Only when they diverge can an output appear from the filter.

In Figure 3.2, when the clock goes low, one of R1 and R2 may go high just before the other, causing the latch to start to change state, but if the overlap is short the latch may be left in metastability. The NAND gate outputs both start high, but the filter outputs are low because the two p-type transistors are nonconducting

When both gate outputs go to the same metastable level, the filter outputs remain low, and as the metastability resolves, the latch outputs diverge. Only when there is a difference of at least V_t between the gate outputs can the filter output start to rise, so that one output rises when the high output gate is at about: $(V_{dd} + V_t)/2$, and the low output gate is at about $(V_{dd} - V_t)/2$.

3.2 EFFECTS OF FILTERING

The event histograms of latches with metastability filters can be affected by the nature of the filter. When the clock in Figure 3.2 is high either R1 or R2 will be low. When the clock goes low both R1 and R2 go high. By SPICE simulations and measurements on sample circuits it is possible to find the output times for a range of input time differences between R1, and R2. Typical results are shown in the event histograms of Figure 3.3 and Figure 3.4.

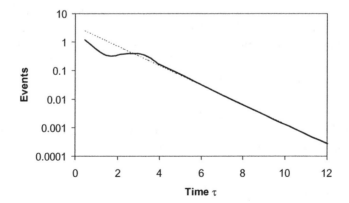

Figure 3.3 Histogram for metastability filter. Reproduced from Figure 6, "Synchronization Circuit Performance" by D.J.Kinniment, A Bystrov, A.V.Yakovlev, published in IEEE Journal of Solid-State Circuits, 37(2), pp. 202–209 © 2002 IEEE.

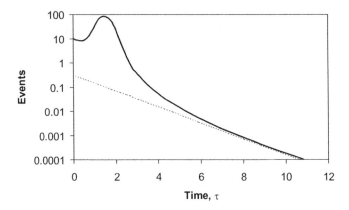

Figure 3.4 Latch with low-threshold output inverters. Reproduced from Figure 7, "Synchronization Circuit Performance" by D.J.Kinniment, A Bystrov, A.V.Yakovlev, published in IEEE Journal of Solid-State Circuits, 37(2), pp. 202–209 © 2002 IEEE.

Figure 3.3 shows the effect of the circuit with a filter of the type shown in Figure 3.2, where the gate outputs both start high, and one of them must go more than $V_t/2$ below metastability to give an output. Output times in this figure are measured as the elapsed time in multiples of τ after the last input goes high. In Figure 3.3 the initial slope is only slightly faster than the final slope, but the effect is more pronounced in Figure 3.4 where the outputs are taken from low-threshold inverters with transistors sized the same as those in the filter. The threshold in the low-threshold inverters is about $V_{dd}/30$ below the metastable level.

The effect of the low-threshold inverter on the early part of the slope of Figure 3.4 can be compared with that of Figure 2.17 (low start) where the early events have been delayed because outputs both started high, and one has to fall to the threshold level before the inverter output goes high. In the long term this effect is negligible, so the histogram approaches the dotted trend line. In Figure 3.3 the slope of the trend line is slower because the loading on the latch output is greater, but the increase in delay applies equally to early and late events, so the difference between initial and final slope is not so pronounced. Though the circuit is slower, it is much less susceptible to noise, since the threshold detecting the resolution of metastability is much further away.

3.3   THE JAMB LATCH

In most applications it is important to reduce the failure rate to a very low value, and therefore the value of τ should be as low as possible.

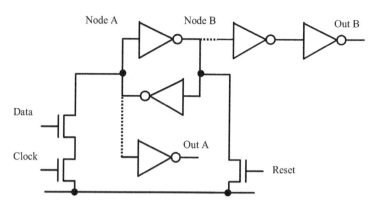

Figure 3.5 Jamb latch.

Circuits called Jamb latches, based on cross-coupled inverters rather than gates are used [10,11] because inverters have a higher gain, and less capacitance, than gates. A schematic of a Jamb latch with the output taken from two places is shown in Figure 3.5.

In the Jamb latch, the cross-coupled inverters are reset by pulling node B to ground, and then set if the data and clock are both high, by pulling node A to ground. The output inverter on node A has a low threshold, and buffers the node from the effects of the output, and protects the rest of the system from the effects of metastability since its output cannot go high until node A goes fully low. It is also possible to take an output from node B, but here there must be two inverters so that the output is the same phase as the input, and two similar latches can be used in a synchronizer. The first inverter has to have a high threshold to buffer the output from metastability.

In a normal CMOS inverter, the p-type devices have a width twice the n-type. This is because it is important to make the timing of the low to high transition on the output the same as the high to low transition. To make the pull-up current from a p-type CMOS transistor the same as the pull-down current from an n-type transistor, the width of the p-type has to be twice the n-type. The situation in metastability is different. Here the transconductance of the inverter is the sum of the transconductance G, of both p-type and n-types, and the capacitance C is also the sum of both devices. Thus the optimum value of $\tau = C/G$ is found when the p-type is much smaller than the n-type, ideally zero. In practice the variation in τ with p-type width is not large, and it is necessary to have a p-type pull-up transistor to hold the latch state so a 1:1 ratio between p and n widths is usually used. For correct operation,

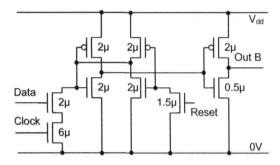

Figure 3.6   Jamb latch B transistor widths.

reset, data and clock transistors must all be made wide enough, when compared to the inverter devices, to ensure that the nodes are properly pulled down during set and reset. Typically, this means that the reset transistor has a similar width to the p-type transistors in the flip-flop, and the data transistor is wider because it is in series with the clock. Typical transistor widths are shown in Figure 3.6.

Metastability occurs if the overlap of data and clock is at a critical value which causes node A to be pulled down below the metastability level, but node B has not yet risen to that level. This can be seen in Figure 3.7, which was produced by simulation, where the data goes high

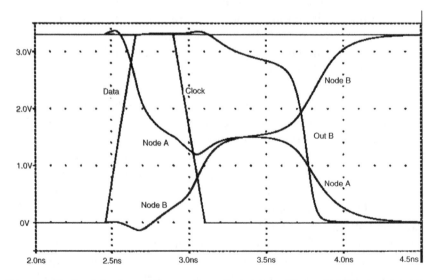

Figure 3.7   Jamb latch waveforms. Reproduced from Figure 9, "Synchronization Circuit Performance" by D.J.Kinniment, A Bystrov, A.V.Yakovlev, published in IEEE Journal of Solid-State Circuits, 37(2), pp. 202–209 © 2002 IEEE.

at about 2.55 ns while the clock is high, then node A falls to about 1.1 V, while node B rises to about 0.8 V. When the clock input goes low at 3 ns both node A and node B become metastable at about 1.5 V, and the output can be taken from node B with an inverter whose threshold is 0.1V higher than the metastable level.

If the latch state is to be observed at node A, we must use a low-threshold inverter between node A and the output because that node is going low. Only one output inverter is actually connected to either node A or node B to avoid loading the latch too much, and in that inverter the transistor widths are minimized. When the node B output is used the inverter threshold and node B start voltage are on different sides of the metastable level so the events histogram should correspond to the high-start, low-threshold curve of Figure 2.17, and in the second (node A output), they are on the same side so this corresponds to the high-start, high-threshold curve. When the clock and data overlap, both nodes always start from below the metastable level. To record an output from the node A Jamb latch a low-threshold inverter is used, so that when node goes fully low, there is a high output. This node is quite sensitive to small changes in the clock data overlap, because it is already below the metastable level and does not have far to move. This means you get a bigger than expected change in output time for a given overlap change, and therefore there are fewer events that can give an output within a fixed output time slot.

The inverter on node B needs to have a high threshold to record an output change, and it is less sensitive to a given change in clock data overlap than would be expected, because the node is low already, and needs to go high to give the low going final output. Turning that into the number of events that fit into an output time slot, there are more than normal because a bigger change in clock data overlap is needed to span the time slot.

Histograms produced from simulations of both node A and node B circuits are shown in Figure 3.8. The node A circuit has both start and threshold on the same side of the metastable level, and is therefore equivalent to the high-start high-threshold case of Figure 2.17, whereas for Node B the start and threshold are on opposite sides, and so are equivalent to the low-start case.

3.3.1   Jamb Latch Flip-flop

A master–slave flip-flop built from two Jamb latches may have the latch output taken either from node A or node B. If node A is used, a single

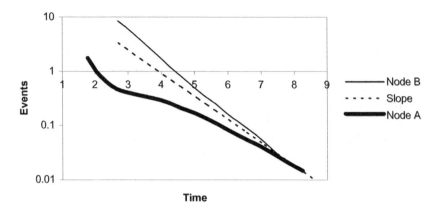

Figure 3.8 Jamb latch histograms. Reproduced from Figure 10, "Synchronization Circuit Performance" by D.J.Kinniment, A Bystrov, A.V.Yakovlev, published in IEEE Journal of Solid-State Circuits, 37(2), pp. 202–209 © 2002 IEEE.

inverter with a low threshold provides a high output when the data is high and the latch is clocked.

For large input times the two latch outputs in Figure 3.8 have a shorter delay for node A than for node B, for example in a simulation of a typical 0.18 μ process when the input is 113 ps before the clock, the output after one inverter is 97 ps after the clock for Out A, a total delay of $T_d = 210$ ps and 147 ps after for node B so $T_d = 260$ ps. However when the input is less than 10 ps, the delays are very similar. Intuitively, it is obvious that the large signal delay from the input through to the first Out B inverter path has one more inverters than the Out A path, and will be approximately one inverter delay longer. The simulation shows that when the clock goes high and the latch is metastable, there is very little voltage difference between nodes A and B, so the two delays are similar when metastability is resolved.

By plotting input times against output times we can deduce the values of τ, T_{w0} and T_d from the simulations

$$\text{For Node A } \tau = 40 \text{ ps}, T_{w0} = 20\,000\,\text{ps and } T_d = 201\,\text{ps}$$

$$\text{For Node B } \tau = 40\,\text{ps}, T_{w0} = 20\,000\,\text{ps and } T_d = 250\,\text{ps}$$

Here we are using T_{w0} which is the intercept on the y-axis at for output times measured from the clock. T_w is the intercept for the typical output time of around 250 ps, and in this case is

$$T_w = 20\,000.e^{-\frac{250}{40}} = 38.6 \text{ ps}$$

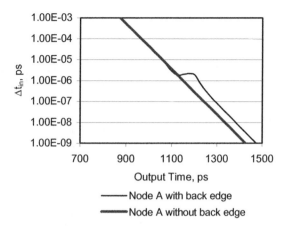

Figure 3.9 Jamb latch Out A back edge offset.

The back edge offset (Equation 2.25), for Out A is therefore, 47 ps and for Out B, −1 ps. 47 ps is similar to the delay expected in an inverter in this technology. Further simulation for a master–slave flip-flop using Node A and with a clock back edge at 1 ns gives the input output characteristic of Figure 3.9.

The difference between T_d and the projection of the deep metastability slope τ, back to the point where $\Delta t_{in} = \tau$ gives an estimate of the offset.

3.4 LOW COUPLING LATCH

The value of τ in these circuits is determined by the drive capability of the inverters, and the capacitive loading on the nodes. To reduce this loading, the output inverters should have small geometry, but the set and reset drive transistors in the Jamb latch cannot be reduced below a certain size, or the circuit will not function correctly. It is possible to overcome this problem by switching the latch between an inactive (no-gain) and an active (high-gain) state. As the device moves between the two states, only a small drive is necessary to cause the output to switch one way or the other, and if this drive is small, it can be maintained in the fully active state without switching the output further.

Figure 3.10 shows a circuit based on this principle, in which the latch is activated by the low to high transition of the clock, and one of the B_1 and B_0 nodes goes low, giving a high output on V_1 if data is high before the clock. Because they don't need to overcome a fully conducting

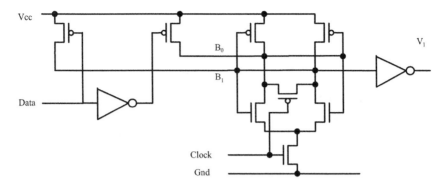

Figure 3.10 Low-input coupling latch. Reproduced from Figure 12, "Synchronization Circuit Performance" by D.J.Kinniment, A Bystrov, A.V.Yakovlev, published in IEEE Journal of Solid-State Circuits, 37(2), pp. 202–209 © 2002 IEEE.

n-type transitor, the *p*-type data drive transistors need to be less than one-quarter the size of those in the Jamb latch, and so load the latch is much less.

The simulated performance of the circuit is shown in Figure 3.11. The output takes at least 5τ from the clock, where the Jamb latch takes only 2τ. This is because the internal bistable starts from near the metastable point when the clock goes high, and only has a very low drive to bias it in one direction or the other, where the Jamb latch is driven very hard by any clock data overlap. The trade off is that τ is only 80% of the equivalent Jamb latch. Because τ is lower, after a resolution time of 20τ it has caught up with the Jamb latch, and the number of events at 30τ is

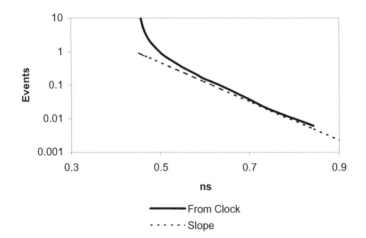

Figure 3.11 Low-coupling latch characteristics.

significantly less. The trade-off is that τ has been made a little smaller, at the expense of T_w, which is significantly bigger, and hence the delays are longer at first. The value of τ in this circuit is lower because loading effects from the set and reset mechanism are lower.

3.5   THE Q-FLOP

In some systems it is useful to know when metastability has resolved. In order to achieve this we can use a metastability filter in a circuit whose ouputs are the same in metastability, and also when the clock is low. A circuit that has these properties is shown in Figure 3.12 This circuit, sometimes known as a Q-flop [12], ensures that half-levels will not appear at the output, and any uncertainty due to metastability appears as a variable delay time from the clock to the output rather than a half-level. When the clock is low, both outputs are low because both Q_L and Q_{Lbar} are high. When the clock goes high the Q-flop can take up the state of the D input at that time, and then becomes opaque to D changes, but it may become metastable if the D input changes near to the clock edge time.

In Figure 3.12 only clean output level transitions occur, because the output is low during metastability, and only goes high when there is a significant voltage difference between Q_L and Q_{Lbar}. Equation (2.5) shows that the transition between a voltage difference of V_1 and V_2 takes a relatively short time, $\tau \ln (V_2/V_1)$, so when the voltage between Q_L and Q_{Lbar} is sufficient to produce an output (more than the p-type threshold V_t) the transition to a full output of V_{dd} will take less than

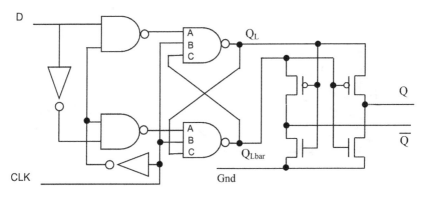

Figure 3.12   Q-flop.

2τ. This voltage difference is also much greater than any latch internal noise voltage, so when the Q_L signal leaves the metastable region to make a high transition, it happens in a bounded time and cannot return low. On the other hand, the time at which this transition takes place cannot be determined, and may be unbounded.

3.6 THE MUTEX

An important component of asynchronous systems is the mutual exclusion element. Its function is to control access to a single resource from two or more independent requests. If access to one request is granted, the others must be excluded until the action required is completed. A simple application is the control of access to a single memory from two processors. Requests from each processor to read from, or write to a particular memory location must not be interrupted by the other, and if the requests are entirely independent, they may occur nearly simultaneously. The design of arbiters, which decide which request to grant are discussed more fully in Part C, but almost all are based on the use of a MUTEX circuit which must distinguish between two closely spaced requests, and metastability is inevitable when the time between requests is very close.

Most MUTEX circuits are based on a set reset latch model in which the latch is made from cross-coupled gates, and is followed by a filter circuit, which prevents metastable levels reaching the following circuits, and also has the function of signalling when the decision is complete. Figure 3.13 shows how this can be done with arrangement similar to Figure 3.1 in which the outputs of the RS latch are fed into two low-threshold inverters.

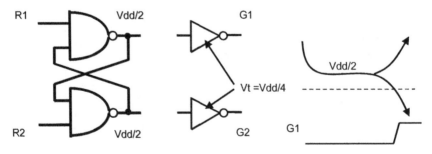

Figure 3.13 MUTEX.

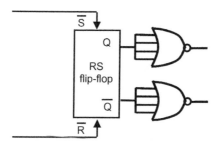

Figure 3.14 FPGA MUTEX with low-threshold inverters.

When both request inputs, R1 and R2, of the MUTEX are low, both of the latch outputs are high. In this case both inverter outputs, G1 and G2, act as the grant outputs and are low; this signals that no request has been granted. If only one request goes high, the corresponding grant will go high, and any subsequent request on the other input will be locked out. If both requests are made within a short time of each other, the first request will normally be granted, but if both grants go high at almost the same time the latch may be left in a metastable state. In metastability both grant outputs are low because the latch metastable level is higher than the output inverter thresholds, shown here as a dotted line. Only when one latch output goes fully low can the corresponding grant go high. It is important to note that a high grant output then indicates the end of metastability in the latch and the completion of the MUTEX decision.

It is more difficult to design a MUTEX in a standard cell or FPGA environment because the transistor geometries cannot be altered to produce a shifted threshold, but a similar effect can be obtained by paralleling the inputs on a multi-input NOR gate to lower the threshold as in Figure 3.14, or using a multi-input NAND gate to raise the threshold. An FPGA MUTEX should also have the flip-flop implemented as a single RS device to avoid the possibility of the oscillation problem described earlier.

Once the latch output voltage has fallen far enough below $V_{dd}/2$ to cause one of the grant outputs to start to go high it does not take long for it to fall all the way to a defined low level. Since the output voltage trajectory of a metastable latch is given by Equation (2.5), then the time taken for the trajectory to go from a threshold level, V_{th} below metastability, to a low logic level V_{low}, is

$$t = \tau \ln\left(\frac{V_{low}}{V_{th}}\right) \qquad (3.1)$$

This time is a constant, depending only on circuit characteristics, so the timing of the end of metastability after the output gate threshold is exceeded is fixed, and the grant cannot return to a low level.

Typically MUTEX circuits are used in situations where the time between the requests R1 and R2 can have any value, and the probability of all separation times is the same. We can assume therefore that the probability of all the values of K_b between 0 and V_{dd} is also the same. Now the average extra time required for resolution of metastability can be found by using Equation (2.5), where

$$V_{dd} = K_b e^{\frac{t}{\tau}}$$

and averaging t over all values of K_b

$$t_{average} = \int_0^{V_{dd}} t \, \mathrm{d}K_b = \tau \tag{3.2}$$

From this it is possible to see that on, average the time penalty imposed by a MUTEX is quite short, at the normal propagation delay plus τ. While there is a finite probability of long resolution times, the probability of times significantly longer than τ decreases exponentially with time, and the probability of an infinite decision time is zero.

3.7   ROBUST SYNCHRONIZER

One of the problems of synchronizers in submicron technology is that latches using cross-coupled inverters do not perform well at low voltages and low temperatures. Since the value of τ depends on the small signal parameters of the inverters in the latch it is more sensitive to power supply variations than the large signal propagation delay expected when an inverter is used normally. This is because the conductance of both the p- and n-type devices can become very low when the gate voltage approaches the transistor threshold voltage V_T, and consequently C/G can become very high. As V_{dd} reduces in submicron processes, and V_T increases, the problem of increased τ and therefore greatly increased synchronization time gets worse. Typical plots of τ against V_{dd} for a $0.18\,\mu$ process are shown in Figure 3.15. It can be observed from this figure that τ increases with V_{dd} decreasing and the reduction in speed becomes quite rapid where V_{dd} approaches the sum of thresholds of p- and n-type transistors so that the value of τ is more

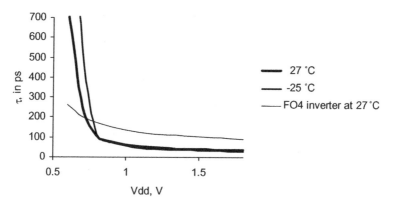

Figure 3.15 Jamb latch τ vs V_{dd}.

than doubled at a V_{dd} of 0.9 V, and more than an order of magnitude higher at 0.7 V, $-25°C$. For comparison the typical large signal inverter delay with a fan out of four (FO4) in this technology is shown. This demonstrates τ is likely to track the processor logic delay rather poorly, making design difficult.

The increase in τ can have a very important effect on reliability. For example, a synchronizer in a system where a time equivalent to 30τ has been allowed might give a synchronizer *MTBF* of 10 000 years. A 33% increase for τ, in this synchronizer will cause the time t_p fall to an equivalent of only 20τ. As a result the *MTBF* drops by a factor of e^{-10} from 10 000 years, to less than 6 months. It is very important, that worst-case variations of all parameters, such as process fluctuations, temperature, and power supply are taken into account in any estimate of τ to ensure that the reliability of the system under all operating conditions is as expected, and circuits are needed that are robust to variations of process, voltage and temperature.

One way of improving the value of τ is to increase the current in the transistors by increasing all transistor widths, but this will also increases power dissipation. In order to estimate the average energy used during metastability, we will assume that the average metastability time is τ. As the transistor width increases, the total switching energy increases in proportion but τ only decreases slowly as transistor sizes increase, and reaches a limit at around 30 ps in a 0.18μ process. While τ can be optimized for conventional circuits, sensitivity to PVT variation remains a problem.

An improved synchronizer circuit [23] that is much less sensitive to power supply variations is shown in Figure 3.16.

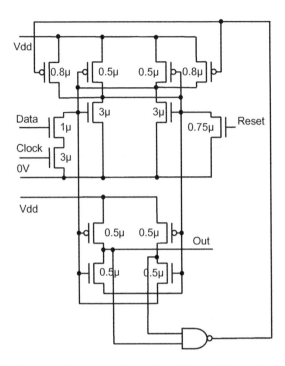

Figure 3.16 Robust synchronizer. Reproduced from Figure 2, "A Robust Synchronizer Circuit", by J Zhou, D.J. Kinniment, G Russell, and A Yakovlev which appeared in Proc. ISVLSI'06, March 2006, pp. 442–443 © 2006 IEEE.

This circuit is essentially a modified Jamb latch where two 0.8 μ p-type load transistors are switched on when the latch is metastable so as to maintain sufficient current to keep the total transconductance high even at supply voltages less than the sum of thresholds of the p- and n-type transistors. Two 0.5 μ feedback p-types are added in order to maintain the state of the latch when the main 0.8 μ p-type loads are turned off. Because of these additional feedback p-types, the main p-types need only to be switched on during metastability, and the total power consumption is not excessive. In the implementation of Figure 3.16 a metastability filter is used to produce the synchronizer output signals, which can only go low if the two nodes have a significantly different voltage. The outputs from the metastability filter are both high immediately after switching, and are then fed into a NAND gate to produce the control signal for the gates of two main p-types. In this circuit, the main p-types are off when the circuit is not switching, operating like a conventional Jamb latch, but at lower power, then when the circuit enters metastability the p-types are switched on to allow fast switching. The main output is taken

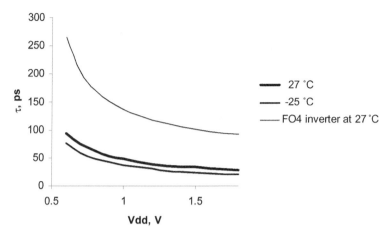

Figure 3.17   Robust synchronizer τ vs V_{dd}.

from the metastability filter, again to avoid any metastable levels being presented to following circuits. Now there is no need for the feedback p-types to be large, so set and reset can also be small. The optimum transistor sizes for the improved synchronizer are shown in Figure 3.16, and the resultant τ at V_{dd} of 1.8 V is as low as 27.1 ps because the main transconductance is provided by large n-type devices and because there are two additional p-types contributing to the gain. It also operates well at 0.6 V V_{dd} and $-25°C$, because it does not rely on any series p- and n-type transistors being both switched on by the same gate voltage.

The relationship between τ and V_{dd} for the improved synchronizer is shown in Figure 3.17.

The switching energy for this circuit is 20% higher than a conventional Jamb latch. At the same time as maintaining a low value of τ, the ratio between τ and FO4 is much more constant at around 1:3 over a wide range of V_{dd} and temperature.

3.8   THE TRI-FLOP

Multistable (as opposed to two state) flip-flops have also been proposed for synchronization and arbitration, and may have an advantage in some situations, however implementations can be unreliable. Figure 3.18 shows a tristable circuit, in which only one of the y outputs can be low when all the r inputs are high. Multi input MUTEX circuits can be made from trees of two or three input circuits an in certain circumstances a

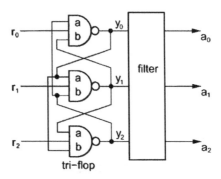

Figure 3.18 Tristable flip-flop. Reproduced from Figure 1, "Analysis of the oscillation problem in tri-flops", by O. Maevsky, D.J. Kinniment, A.Yakovlev, A. Bystrov which appeared in Proc. ISCAS'02, Scottsdale, Arizona, May 2002, IEEE, volume I, pp. 381–384 © 2002 IEEE.

tree based on three input MUTEXes may be lower power and higher performance than those based on two input MUTEXes. Unfortunately without careful design they can oscillate when two, or all three request signals r0, r1, and r2 arrive simultaneously.

The small signal model for a tristable can be described by three first-order differential equations

$$-C_x \frac{dV_x}{dt} = G_x V_x + G_{xy} V_y + G_{xz} V_z$$

$$-C_y \frac{dV_y}{dt} = G_{yx} V_x + G_y V_y + G_{yz} V_z \qquad (3.3)$$

$$-C_z \frac{dV_z}{dt} = G_{zx} V_x + G_{zy} V_y + G_z V_z$$

where C_x, C_y, C_z represent the output capacitances of gates x, y, and z, and G_x, G_y, G_z their output conductances. Similarly, G_{xy}, G_{xz}, etc. represent the input transconductances of gate x, etc. from the inputs connected to gates y and z . If $C_x = C_y = C_z = C$, $G_x = G_y = G_z = G$ and $G_{xy} = G_{yz} = G_{zx} = G_a$, $G_{xz} = G_{zy} = G_{yx} = G_b$, a solution can be found where:

$$V = K_a e^{-2\frac{t}{\tau}} \begin{pmatrix} 1 \\ 1 \\ 1 \end{pmatrix} + K_b e^{\frac{t}{\tau}} \begin{pmatrix} \sin(\omega t + \varphi_0) \\ \sin\left(\omega t + \varphi_0 - \frac{2\pi}{3}\right) \\ \sin\left(\omega t + \varphi_0 + \frac{2\pi}{3}\right) \end{pmatrix} \qquad (3.4)$$

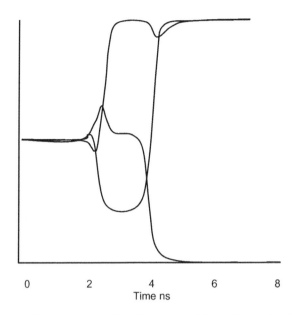

0   2   4   6   8
Time ns

Figure 3.19   Oscillation in a tri-flop. Reproduced from Figure 2, "Analysis of the oscillation problem in tri-flops", by O. Maevsky, D.J. Kinniment, A.Yakovlev, A. Bystrov which appeared in Proc. ISCAS'02, Scottsdale, Arizona, May 2002, IEEE, volume I, pp. 381–384 © 2002 IEEE.

Here

$$\omega = \frac{\sqrt{3}}{2} \frac{|G_a - G_b|}{C}$$

If the two input transconductances, G_a, G_b are consistently different, metastability may result in oscillations on the outputs, which cannot be filtered out [13]. Taking this to the extreme where $G_b = 0$, but G_a is not, the 'b' inputs to the gates would be open circuit, and we would have a ring of three oscillator. This oscillatory metastability is clearly shown in Figure 3.19 where the tri-flop was implemented using 0.6μ CMOS 'symmetrical' NAND gates in which the width of the input transistors was twice that of the 'b' input. Realistic worst-case tolerances of technological parameters also give a similar, though less vivid effects.

If the input transistor geometries are made identical, ω becomes zero, and there is no oscillation, but small variations in the transistor critical length dimensions could easily tip the circuit into instability, and a more robust solution can be found by examining the roots of the characteristic equation

$$p^3 + ap^2 + bp + c = 0 \tag{3.5}$$

where

$$a = \frac{G_x}{C} + \frac{G_y}{C} + \frac{G_z}{C}$$

$$b = \left(\frac{G_x G_y}{C^2} - \frac{G_{xy} G_{yx}}{C^2} \right) + \left(\frac{G_x G_z}{C^2} - \frac{G_{xz} G_{zx}}{C^2} \right) - \left(\frac{G_y G_z}{C^2} - \frac{G_{yz} G_{zy}}{C^2} \right)$$

$$c = \frac{G_x G_y G_z}{C^3} + \frac{G_{xy} G_{yx} G_{xz}}{C^3} + \frac{G_{xz} G_{xy} G_{yx}}{C^3} - \frac{G_x G_{yz} G_{zy}}{C^3} - \frac{G_y G_{xz} G_{zx}}{C^3} - \frac{G_z G_{xy} G_{yx}}{C^3}$$

Imaginary roots will result in an oscillatory solution to the system of differential equations, but stability, with a tolerance of 10% variations in transconductance can be assured [14], by making:

$$\frac{G_x}{C} = \frac{G_y}{C} = \frac{G_z}{C} = \frac{G_{xz}}{C} = \frac{G_{zx}}{C}$$

$$= 1.1 \frac{G_{xy}}{C} = 1.1 \frac{G_{yx}}{C}$$

$$= 0.9 \frac{G_{yz}}{C} = 0.9 \frac{G_{zy}}{C} \qquad (3.6)$$

The potential gain from using a tri-flop, however, is not likely to be worth the effort of ensuring the correct transconductance conditions, and networks of two-way MUTEXs will generally be more advantageous when building multiway arbiters.

4

Noise and its Effects

4.1 NOISE

The effect of noise is that the output of a synchronizer becomes nondeterministic with the input close to the balance point, and an individual output time no longer depends primarily on the inputs. This only occurs with very small timing differences, and to understand the effect of a noise signal measured in millivolts it is first necessary to find the relationship between voltage displacement and time. By measuring the initial voltage difference between the two latch outputs in a simulated Jamb latch resulting from very small changes in the overlap of clock and data, the trade-off θ, between time overlaps and equivalent voltage difference can be found.

Actual measurement of the noise is best done using a small DC bias to create a given voltage difference between the nodes rather that a time difference between data and clock signals because external jitter and noise can be filtered out from the DC bias, where it cannot be easily removed from high-bandwidth inputs.

The measurements can be made with the circuit of Figure 4.1. A constant current is fed into one node of a flip-flop and out of the other, so that a small variable bias is imposed between the nodes. Then the flip-flop is activated by continuously clocking while the bias is slowly varied.

When V_0 is very close to V_1 at the time the clock goes high, the flip-flop output is determined mainly by thermal noise, since the RMS noise voltage on these nodes is greater than the offset due to the input current i. Under these circumstances the random nature of the output can be

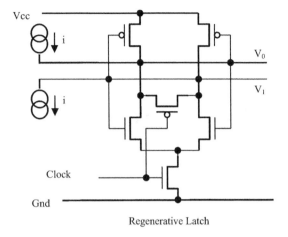

Regenerative Latch

Figure 4.1   Noise measurement circuit.

clearly observed, and as the input current changes from negative through zero to positive the proportion of V_1 high outputs goes from zero to 100%. Plotting the change in this proportion for a given input change against the actual input for one sample of the fabricated devices gives a graph like that of Figure 4.2, where the points measured are compared with an error function curve with an equivalent RMS noise value.

The origin of noise in CMOS transistors is fluctuations in the drain current due to quantum mechanical effects. These thermal fluctuations

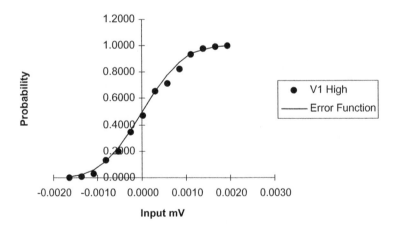

Figure 4.2   Probability of V_1 high. Reproduced from Figure 15, "Synchronization Circuit Performance" by D.J.Kinniment, A Bystrov, A.V.Yakovlev, published in IEEE Journal of Solid-State Circuits, 37(2), pp. 202–209 © 2002 IEEE.

give rise to drain current noise and gate noise. Van der Ziel [15] gives drain current noise as

$$i_{nd}^2 = 4kT\gamma g_{d0}\Delta f \qquad (4.1)$$

where g_{d0} is the drain–source conductance, and the parameter γ has a value typically between 2 and 3. The equivalent gate noise voltage is approximately

$$e_{ng}^2 = \frac{4kT\delta\Delta f}{5g_{d0}} \qquad (4.2)$$

where δ is typically 5–6. The amplitude of the noise is related to the conductance and the bandwidth over which it is measured. In the latch circuit noise at nodes V_1 and V_0 has a bandwidth limited by the capacitance C to:

$$\Delta f = \frac{g_d}{4C} \qquad (4.3)$$

By adding the two noise contributions due to Equations (4.1) and (4.2), then putting $C = \dfrac{g_d}{4\Delta f}$ a good approximation to the gate noise at each node is obtained as:

$$e_n = \sqrt{\frac{3kT}{C}} \qquad (4.4)$$

Where T is the absolute temperature, and k is Boltzmann's constant, 1.38 $\times$ 10^{-23}. This is 0.5–0.6 mV on each node for a 0.6 μ process, but between nodes V_1 and V_0 it is $\sqrt{2}$ greater, or about 0.8 mV. A measurement of approximately 0.8 mV RMS corresponds to a time difference of about 0.1 ps at the inputs of a Jamb latch. It is interesting to note that as the dimensions of the device reduce, the effects of noise become greater, with a 60 nm process giving three times the noise voltage. If the supply voltage is also reduced by a factor of three, the effects of noise are also greater, so that the time differences at which indeterminacy starts are around 0.3 ps unless the transistor W/L ratios are increased to give a higher value of C.

Intuitively it seems that a latch cannot remain in metastability in the presence of random noise, and because it gets knocked off balance, the metastability time will be shortened. In a typical synchronizer application, this is not the case. Couranz and Wann [16], have demonstrated both theoretically and experimentally that for a uniform distribution of initial condition voltages, as would be the situation for the histograms presented here, the probability of escape from metastability with time does not change with the addition of noise, but only if we assume a uniform distribution of initial conditions. For each noise contribution that moves a trajectory away from metastability, there will, on average, be another compensating noise contribution that moves a trajectory towards metastability.

4.2   EFFECT OF NOISE ON A SYNCHRONIZER

To show how this happens, let us assume a normal distribution of noise voltages such that the probability of a noise voltage being within dv_1 of v_1 is

$$P_1(v_1) = \frac{1}{e_n\sqrt{2\pi}}e^{\frac{-v_1^2}{2e_n^2}}dv_1 \tag{4.5}$$

Here the RMS noise voltage is e_n, and the probability of the noise voltage being somewhere between $-\infty$ and $+\infty$ is

$$\int_{-\infty}^{\infty}\frac{1}{e_n\sqrt{2\pi}}e^{\frac{-v_1^2}{2e_n^2}}dv_1 = 1 \tag{4.6}$$

Further, we will assume for a normal synchronizer that the probability of a particular initial difference K_b being within dv of v is

$$P_0(v) = \frac{dv}{V} \tag{4.7}$$

where V is the range of voltages over which the probability is constant. If the probability density of the initial voltage difference is constant, then the probability density of the trajectories at any time is also constant. This can be seen by observing that the trajectories cross both the $X = 0.1$ line and the $X = 1.1$ line at different, but regular spacing in Figure 4.3.

We convolve these two to get the resulting probability density of initial differences at the latch nodes.

$$P(v) = \int_{-\infty}^{\infty}P_1(v_1)P_0(v-v_1)dv_1dv \tag{4.8}$$

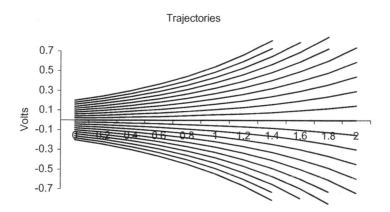

Figure 4.3 Uniformly spaced trajectories.

In a normal synchronizer $P_0(v - v_1)$ is a constant over a range much wider than the noise voltage, $(V \gg e_n)$, so it can be taken outside the integral

$$P(v) = \frac{1}{V} \int_{-V/2}^{V/2} \frac{1}{e_n\sqrt{2\pi}} e^{\frac{-v_1^2}{2e_n^2}} dv_1 \quad \text{and} \quad \int_{-\infty}^{\infty} \frac{1}{e_n\sqrt{2\pi}} e^{\frac{-v_1^2}{2e_n^2}} dv_1 = 1 \quad (4.9)$$

The final probability density is also constant,

$$P(v) = \frac{dv}{V} \quad (4.10)$$

and the result is nearly the same as the one we started with.

The process of convolution is illustrated in Figure 4.4, where the noise distribution $P_1(v)$, is convolved with the initial difference distribution $P_0(v)$, to produce the result $P(v)$. In fact it does not matter much that we used a Gaussian distribution of noise voltages, any other distribution with negligible probability outside $-V/2$ and $+V/2$ would give a similar result.

4.3 MALICIOUS INPUTS

4.3.1 Synchronous Systems

Sometimes clock and data can become intentionally, or unintentionally locked together so that the data available signal always changes at exactly the same relative time. If this time is the balance point of the latch, the

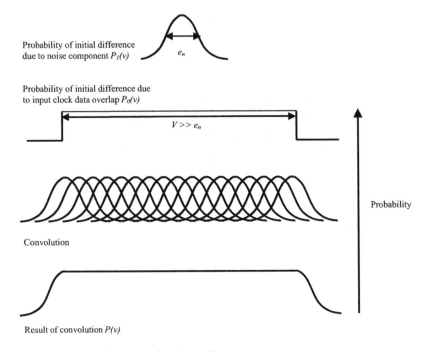

Probability of initial difference
due to noise component $P_1(v)$

e_n

Probability of initial difference due
to input clock data overlap $P_0(v)$

$V \gg e_n$

Probability

Convolution

Result of convolution $P(v)$

Figure 4.4 Convolution with noise voltage.

input is called a malicious input, and metastability times could become very long were it not for the presence of noise.

For a malicious input the initial difference voltage between the nodes is always 0 V, and the situation is very different from a normal synchronizer because $V \ll e_n$, so that if $v = 0$, $P_0(0) = 1$, and if $v \neq 0$ $P_0(v) = 0$. Looking at the two cases separately, when $V_1 = v$

$$P_1(v_1)P_0(v - v_1) = P_1(v) \tag{4.11}$$

Otherwise $P_1(v_1) \, P_0(v - v_1) = 0$

This means that

$$\int_{-\infty}^{\infty} P_1(v_1)P_0(v - v_1)\mathrm{d}v_1 = P_1(v) \qquad \text{or} \qquad P(v) = \frac{1}{e_n\sqrt{2\pi}} e^{\frac{-v^2}{2e_n^2}} \mathrm{d}v \tag{4.12}$$

So the result of adding noise to an initial time difference between clock and data difference that is always equal to the balance point, is to produce a distribution of initial differences equal to the noise distribution. This is the intuitively obvious result shown in Figure 4.5, in which the noise has the effect of knocking the metastability off balance so that it resolves quicker.

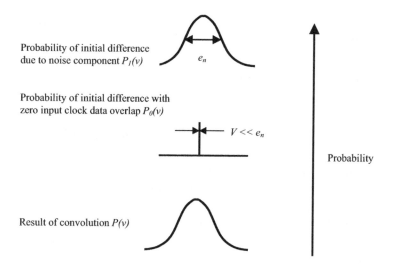

Probability of initial difference
due to noise component $P_I(v)$

e_n

Probability of initial difference with
zero input clock data overlap $P_0(v)$

$V \ll e_n$

Probability

Result of convolution $P(v)$

Figure 4.5 Malicious input.

The effect of noise on the average metastability time is now determined by the probability of the initial difference being near the balance point, so it is possible to compute the *MTBF* for a synchronizer in a typical application where all values of data to clock times are equally probable. This was shown to be

$$MTBF = \frac{e^{\frac{t}{\tau}}}{f_d f_c T_w}$$

in Equation (2.10). For the case where the initial difference is determined solely by noise, the probability of a long metastable time where K_b is less than V_x is given by the number of clocks in time T, multiplied by the probability of a noise voltage less than V_x or

$$\left[f_c T \right] \left[\frac{V_x}{e_n \sqrt{2\pi}} \right] \tag{4.13}$$

This gives

$$MTBF = \frac{e_n \sqrt{2\pi} . e^{\frac{t}{\tau}}}{f_c V_e} \tag{4.14}$$

Comparing the two formulae, (Eqations 2.10 and 4.14) it can be seen that f_d no longer appears in (4.14) because the data is assumed to change at exactly the same time as the clock. T_w is associated with input conditions, which do not now determine the input time, and the noise allows the flip-flop to resolve quicker, larger noise voltages giving longer $MTBF$, and no noise meaning zero $MTBF$.

As an example, if $T_w = 10$ ps, $e_n = 0.8$ mV, and $f_c, f_d = 100$ MHz, a normal synchronizer would have an $MTBF$ of $10^{-5} e^{\frac{t}{\tau}}$, where one with a malicious input would have an $MTBF$ of $2 \times 10^{-10} e^{\frac{t}{\tau}}$.

To get the reliability for a synchronizer with a malicious input to be the same as the normal synchronizer reliability with a uniform distribution of clock-data overlaps we would need a $\tau. \ln[10^{-5}/2.10^{-10}]$ $= 11\tau$, longer synchronization time.

A value of $e_n = 0.8$ mV, or 0.1 ps time variation, would represent the effects of internal thermal noise only. In practice jitter on the clock or data of a small system might be 4 ps, and in a large system 20 ps clock jitter would not be unusual. If this jitter has a Gaussian distribution it would be sufficient to add

$$\tau\ln\left[\frac{10^{-5}}{80.10^{-10}}\right] = 7.1\tau \quad \text{or} \quad \tau\ln\left[\frac{10^{-5}}{400.10^{-10}}\right] = 5.5\tau \qquad (4.15)$$

to the synchronization time in order to get the same reliability.

4.3.2   Asynchronous Systems

In an asynchronous system arbitration is subject to delay due to meta-stability, but if the relative timing of the competing requests is uniformly distributed as described in Section 3.6, the average additional delay from this cause is only τ. In many systems the requests are not uniformly distributed, for example in a pipelined processor where both instructions and operands generate requests from different parts of the pipeline to the same cache memory. In this case the dynamics of the system may cause both requests to collide frequently, and in the worst case, always at similar times. In this worst-case situation jitter and noise could spread the request spacing over a range of, say, 5 ps rather than exactly zero every time.

The time taken by the MUTEX to resolve can be obtained from Equation (2.7) and is

$$t = \tau\ln\left[\frac{T_w}{\Delta t_{in}}\right] \qquad (4.16)$$

If the time variation due to jitter and noise is T_n, we need to average the response time over the range 0 to T_n, that is:

$$\text{Average time} = \tau \int_0^{T_n} \ln\left(\frac{T_w}{\Delta t_{in}}\right) d\Delta t_{in} \qquad (4.17)$$

The result of averaging is:

$$\text{Average time} = \tau \cdot \left[1 + \ln\left(\frac{T_w}{T_n}\right)\right] \qquad (4.18)$$

So if the noise is as wide or wider than the metastability window, the extra time is still only τ, but if it is much less, for example if T_w is 100 ps, and T_n as low as 5 ps, the average time might be 4τ rather than τ.

5

Metastability Measurements

5.1 CIRCUIT SIMULATION

Metastability is an analog phenomenon, and so the performance of a circuit can be simulated by a circuit simulator such as SPICE. Predicting synchronizer failure rates, however, is difficult because the aim of a synchronizer is to give a very low failure rate, so that the probability of failure after the synchronization time has elapsed is around 10^{-15}. This means that the voltage difference between the two flip-flop nodes when it enters metastability is very small, at around $V_{dd} \times 10^{-15}$. Circuit simulators are software based, and run on digital processors, so represent the voltages and currents in the circuit by means of floating point numbers, which are limited in their range and accuracy. Expressing two voltages of around 1 V to an accuracy of better than 10^{-15} V requires more than the usual word length used in conventional simulators and each increase in synchronization time requires a corresponding increase in the number of bits needed to represent the variables. Other problems are that below 1 mV node difference noise becomes important, the synchronizer response is nondeterministic, and so the results of a deterministic simulation may, or may not be a true representation of the results in practice. Typically the noise voltage dominates below about $V_{dd} \times 10^{-4}$ so below this point it must be assumed that statistically noise has no effect on $MTBF$. Simulation can be carried out in the deterministic region, and can provide a useful indication of the $MTBF$ of a synchronizer for at least the first five decades of $MTBF$. Even this is not straightforward,

---

but techniques have been developed to obtain a useful guide to *MTBF* using simulation alone.

5.1.1   Time Step Control

The resolution time of a latch is largely determined by the amount of time the output spends close to the metastable level. Once it departs significantly from this level it rapidly progresses to either V_{dd} or 0 V. It is very important that this region close to the metastable level is accurately modeled by the simulator.

Most simulators operate by solving the circuit equations in a series of time steps where the capacitor voltages are numerically integrated by the simulator. If the voltages and currents are known at a particular time, and the circuit consists of resistances and capacitances, it is possible to calculate how much the voltages across the capacitors will change over a small period of time, much less than the time constant of any resistor and capacitor combination in the circuit. The assumption is that the current charging or discharging each capacitor will remain constant over the time. The new voltage at the end of the time step will then be equal to the voltage at the beginning of the time step plus the charging effect of the current. More formally

$$V_{t+dt} = V_t + i_t \frac{dt}{C}$$

where V_t is the voltage across a capacitor at time t, i_t is the current into the capacitor, and dt is the time step. New values are then computed for the currents into the capacitors in the next time step by solving the circuit equations assuming the voltages across the capacitors are constant. To maintain accuracy dt must be sufficiently small to keep any voltage changes small so that the value of i_t does not change significantly during dt. If dt is too large accuracy will be lost, but if dt is too small many iterations are needed, and the simulation takes too long. Simulators try to adjust the value of dt as the simulation progresses so that the user does not need to make adjustments. The automatic adjustment of time step can cause problems when metastability is being simulated. The output trajectory for a metastable latch is normally exponential, and is for the majority of the response time very close to the metastable level. Because none of the circuit nodes changes much during this time, the simulation step is automatically increased until the error exceeds some

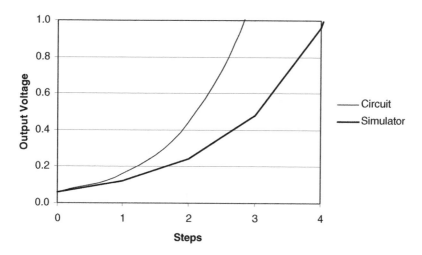

Figure 5.1 Simulation and actual circuit response.

preset bound. The effect of a simulation step that is too long can be seen in Figure 5.1.

Here a positive exponential, such as that which would be seen in an actual circuit is compared with the response computed from a simulation where the time step is too long. Because the flip-flop node voltage difference increases exponentially during the time step and the simulation assumes a linear increase there will be a small, but increasing time lag between the actual response and the simulated response which leads to a large inaccuracy in response time. In Figure 5.1 they are nearly two times steps different by the time the trajectory reaches 1. The problem can be alleviated by setting the time step ceiling (the maximum time step the simulator is allowed to take) to small value less than 1% of the expected value of τ. Too low a value will give very long simulation times, but 1% is usually good enough and by fixing the time step ceiling the variability of the automatic time step will be improved.

5.1.2 Long-term τ

With a small enough time step the main limitation of the simulator is the accuracy of representation of the flip-flop node voltages and the input times of the clock and data signals. These quantities have to be represented by floating point numbers, but it is the difference between

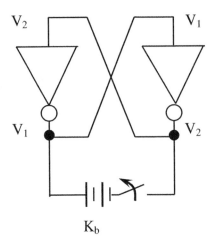

Figure 5.2   Measuring τ with a simulator.

the two that determines the accuracy of the simulation for long resolution times. The ratio between the absolute value of these voltages or times and the smallest difference between the two voltages or two times is usually limited by what can be accommodated in the mantissa of a floating point number. Typically this is around 10^9, well short of the 10^{15} required to accurately model metastability at the synchronization times necessary for adequate $MTBF$. It is also the case that the dynamics of synchronizer circuits can lead to numerical instability in the integration routines so that the accuracy of the simulator is much less than the floating point resolution. Simulation is a useful tool for circuit design, but simulations cannot explore the very low failure probabilities required for real designs.

Rather than simulating long metastability times, is possible to measure the long term value of τ by simulation [10]. This is done by bringing the flip-flop to a starting point, a defined distance from the balance point, and then allowing the metastability to resolve.

In Figure 5.2 two inverters are cross-coupled to make a flip-flop, but both outputs are connected together via a switch and a voltage source. If the switch is closed V_1 and V_2 must have the offset given by the voltage source, so that $V_1 - V_2 = K_b$. Simulating this circuit for sufficiently long for the negative exponential $K_a e^{-t/\tau_a}$ to settle, the switch is then opened. Now V_1 and V_2 start to diverge as the simulation progresses according to the positive exponential $K_b e^{t/\tau_b}$, and the long-term resolution time constant τ_b can be measured as in Figure 5.3. Here, the voltage

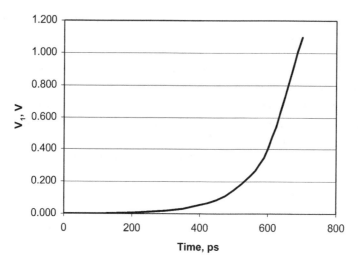

Figure 5.3 Measuring long-term τ.

source is set at 1 mV, and the V_1 output trajectory crosses 1 V at around 700 ps. This means that the value of $τ_b$ is given by

$$τ_b = \frac{700\,\mathrm{ps}}{\ln\left[\dfrac{1\mathrm{V}}{1\mathrm{mV}}\right]}$$

or 101.3 ps. This method can be used for estimating the variation of long term τ with temperature and V_{dd}. Using the value of τ measured in this way, the *MTBF* at quite long synchronization times can also be estimated.

5.1.3 Using Bisection

Another similar method for computing synchronizer failure probabilities [17] is to use a simulator for numerical integration to perform large-signal analysis. This accounts for the nonlinear behaviour of real synchronizer circuits. More accurate small-signal techniques can then characterize behaviours near the balance point. This combination overcomes the limitations of numerical integration in simulators and the technique can be used to understand the transfer of metastable behaviour between synchronizer stages which gives the back edge effect.

This method is based on two observations. First, the size of the window of input events in which the synchronizer fails, Δt_{in} decreases exponentially with the amount of time that the synchronizer has to settle, thus, trajectories corresponding to a slightly early input t_h and slightly late input t_l will be extremely close to each other during the initial part of the simulation where conventional simulation is used, only diverging from each other when the metastable condition is finally resolved. Second, $MTBF$ depends on the time difference between, these events. The exact values of the times are not critical. Calculating the difference directly by a small signal method avoids the 'small difference of large numbers' problem associated with simulation.

Rather than simulating an entire output trajectory, the trajectories are divided into many segments. For each segment two trajectories close to each other are chosen, one of which resolves high and one low. The dynamics of metastable circuits ensures that trajectories that resolve to different logical states will diverge exponentially with time.

Figure 5.4 shows two time segments, each with a pair of trajectories, one which will resolve high and the other low. The first pair correspond to inputs at the slightly early time t_h and the slightly late input t_l. After a time T_1 the difference between the two trajectories, initially $V_{h1} - V_{l1}$ will have diverged by a factor $\alpha = e^{T_1/\tau}$ to $V_{h2} - V_{l2}$. We can compute

$$\alpha = \frac{V_{h2} - V_{h1}}{V_{l2} - V_{l1}} \tag{5.1}$$

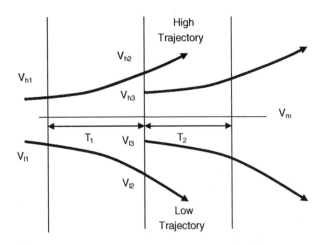

Figure 5.4   Time segments with pairs of trajectories.

from the four output voltages at the beginning and end of the time segment, the balance point

$$V_m = \frac{V_{h2} - \alpha V_{h1}}{1 - \alpha}$$

(5.2)

and the input time t_x corresponding to any starting point V_x, since

$$\frac{t_l - t_x}{t_l - t_h} = \frac{V_{h1} - V_x}{V_{h1} - V_{l1}}$$

(5.3)

We do not need to know the time T_1 and since the calculation is small signal it can be relatively accurate.

Now two new starting points closer to the balance point than V_{h2} and V_{l2} are chosen for the next time segment, these are V_{h3} and V_{l3}. The equivalent starting times for these new points can be found by using Equation (5.3). This procedure is called bisection, and is repeated for each of many time segments. Each new pair of input times will be closer to the balance point than the previous one. Since $MTBF$ can be calculated from the input time interval Δt_{in} using Equation (2.8) the method gives more reliable estimate of $MTBF$ for long synchronization times.

5.2 SYNCHRONIZER FLIP-FLOP TESTING

Measurement of the characteristics of a synchronizer is usually done by using two independent oscillators with slightly different frequencies to provide a spectrum of equally probable set-up and hold times for the flip-flop under test. These two oscillators drive the data and clock inputs of the master–slave flip-flop under test and the basic circuit is shown in Figure 5.5

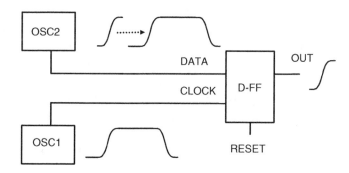

Figure 5.5 Two-oscillator method.

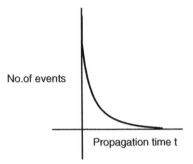

Figure 5.6   Event histogram.

Because the two oscillators have a similar (but slightly different) frequency, the time between the rising edge of the data signal and the rising edge of the clock slowly moves from half a cycle earlier, $\Delta t_m = 1/2f_c$, to half a cycle later, $\Delta t_m = -1/2f_c$. The probability of the time difference Δt_{in} having any particular value between Δt_{in} and $\Delta t_{in} + d\Delta t_{in}$ is always the same in this experiment at $d\Delta t_{in}\, fc$. The characteristics of the flip-flop are measured by counting the number of times the output response time is within the interval t and $t + \delta t$ for each value of t and each count is called an event. Typical results showing events vs time t, are shown in Figure 5.6.

In this figure the number of events is plotted against propagation time t. The curve has a negative exponential form as described by Equation (2.13)

$$\frac{\delta t}{\tau}\left[f_d f_c T T_w e^{\frac{-t}{\tau}}\right]$$

so there are many more events close to the y-axis than for larger values of t away from the y-axis. The main interest in measuring metastability is to estimate the number of times t might exceed the resolution time allowed in a real system, causing the system to fail. The number of such events should be very low for times near the resolution time allowed, but in the measurement the very small number of late events can be obscured by the much more frequent normal flip-flop propagation times near the y-axis in Figure 5.6. This is because events are usually recorded on a digital oscilloscope which may not be able record all the events when they are closely spaced in time.

To avoid that problem, the two oscillators are set to be close in frequency, so that the data oscillator might have a period of 99.9 ns and the clock 100 ns. As the data edge gets close to the rising clock

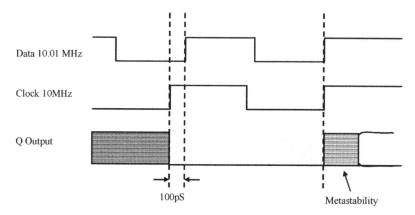

Data 10.01 MHz

Clock 10MHz

Q Output

100pS

Metastability

Figure 5.7 Two-oscillator metastability measurement. Reproduced from Figure 1, "Measuring Deep Metastability", by D.J. Kinniment, K Heron, and G Russell which appeared in Proc. ASYNC'06, Grenoble, France, March 2006, pp. 21–1 © 2006 IEEE.

edge there may, or may not be a change in the output Q, depending on whether the D input had a different value on the previous clock edge. Figure 5.7 shows the situation when the D input is low on the first clock edge and then goes high very close to the next edge, causing a change in the Q output.

This change in Q is used to trigger a digital oscilloscope and the clock edge which caused it is recorded by the oscilloscope. The clock trace is only seen if the D input is different on successive clock edges and then, if the frequencies of the two oscillators are close the data input will have changed close to the second clock edge with a high probability of causing metastability to occur. In this way only those events where clock and data overlap by less than the difference between the two oscillator periods (100 − 99.9 ns) are seen. This is because they are the only events which generate a change in Q. These events are collected by the oscilloscope which can present them as a histogram of the number of events resolved between t and $t + dt$.

The histogram of a Schottky TTL flip-flop in which the data has been collected in this way is shown in Figure 5.8 and other examples appear in [18] and [19]. In Figure 5.8 the x-axis represents time, with the triggering Q output as the reference. When an event is detected by the Q output change, the corresponding clock rising edge time is recorded. Since the clock occurs before the Q output increasing metastability time is shown from right to left in the resulting histogram. The y-axis is the number of events, with a peak of 80 000 at 20.5 ns and the decreasing number of events recorded shows in the discrete levels of 0, 1 and 2 events at around 17 ns.

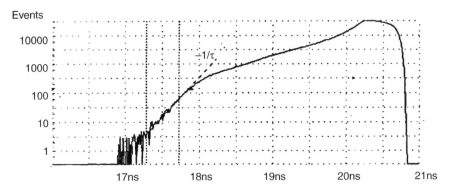

Figure 5.8   Event histogram for a Schottky TTL flip-flop. Reproduced from Figure 2, "Measuring Deep Metastability", by D.J. Kinniment, K Heron, and G Russell which appeared in Proc. ASYNC'06, Grenoble, France, March 2006, pp. 21–1 © 2006 IEEE.

In Figure 5.8 events are plotted on a log scale so that the slope $-1/\tau$ can easily be measured.

It is not necessary to use an oscilloscope for data collection, selecting events can be done by an on-chip change detector, and they can be accumulated in counters. One on-chip measurement method for synchronizers is called a late transient detector (LTD) [20] and the circuit is shown in Figure 5.9. The two oscillators drive the data and clock inputs as before, and the aim of the circuit is to observe the output, Q1, of flip-flop #1, which feeds flip-flop #2 after one clock period, and a delayed and inverted version which feeds flip-flop #3. If the outputs of flip-flop #2 and flip-flop #3 are the same, a metastable event is recorded by flip-flop #4, and the number of these events within a given period can be counted. The propagation time through the multiplexer and flip-flop #1 is first established by setting the calibrate input to high. The clock

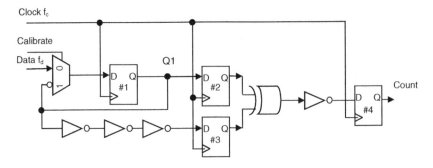

Figure 5.9   Late transient detector.

frequency is then increased until flip-flop #1 fails to divide by 2. At this point the clock period is equal to the normal propagation time t_{c0}. Lowering the frequency until flip-flop #3 responds correctly gives a longer clock period t_c and the difference between the periods, $\delta t = t_c - t_{c0}$, is the time taken by the three inverters. Normally the outputs of flip-flop #2 and flip-flop #3 will be different because of the inversion in the path to flip-flop #3, but if they are the same, it is assumed to be because #3 was metastable. The time difference between #2 and #3 is the interval of δt. Switching to test mode by lowering the calibrate signal allows a clock period t'_c to record events that give response times between $t'_c - t_{c0} - \delta t$ and $t'_c - t_{c0}$ so that an event histogram (number of counts detected within an interval of δt from time $t'_c - t_{c0}$) can be plotted.

One problem with this circuit is the difficulty of controlling the threshold levels of flip-flop #2 and the inverters in relation to the metastable level of flip-flop #1. For example, if the threshold of the first inverter in the chain, feeding flip-flop #3 is slightly higher than the metastable level, the D input of #3 will be high when flip-flop #1 is metastable. If the threshold of #2 is also slightly higher than the metastable level, its output will be low, and because #2 and #3 outputs are the same, the event will not be recorded. On the other hand, if the inverter threshold level is low, the event would be recorded. Another problem is the strong probability of flip-flop #3 being metastable itself at the high frequencies needed to detect events with short metastability times. Thus some of these events may not be correctly recorded.

5.3   RISING AND FALLING EDGES

The late transient detector can detect metastable events due to both rising and falling outputs. Some circuits, for example the Jamb latch described in Section 3.3, are designed so that it is first reset, and then input data is sampled on the next clock cycle. This means that a flip-flop based on Jamb latches can only go metastable when moving from a low to a high output (rising edge). The output inverters of both the node A and node B Jamb latches have their thresholds shifted so that the output transition does not appear until metastability is resolved, and the only uncertainty is in the time taken for the transition to appear at the output. Other flip-flops may not be so carefully designed, and metastability can affect both edges as well as the output level.

How this happens is illustrated in Figure 5.10. The output of the master latch in a flip-flop may start high or low when the set-up and hold times are violated, and it may then become metastable for an

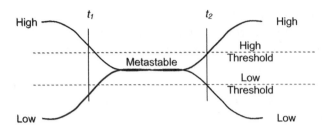

Figure 5.10   Possible master latch outputs.

indeterminate period of time, finally resolving to either a high or low level. The behaviour of the slave output, which is transparent when the clock is high, depends on whether its input threshold is higher or lower than the metastable level of the master. If the threshold is higher, the slave output may change from low to high at t_2, but it also can change high to low at t_1, or even high to low at t_1 and low to high at t_2. If the threshold is lower, the slave output can change from high to low at t_2, low to high at t_1, or low to high at t_1 and high to low at t_2. Thus a two-oscillator experiment can produce rising outputs at t_2 and falling outputs at t_1 for one flip-flop, and for another flip-flop it produces rising outputs at t_1 and falling outputs at t_2.

Figure 5.11 shows how the late transient detector can be modified to count rising and falling edges separately. If Q1 is first low then high, both flip-flop #2 and flip-flop #3 are set high, so a rising edge is counted. If Q1 is high, then low, both flip-flop #2 and flip-flop #3 are set low and a falling edge is counted.

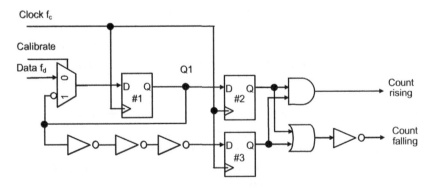

Figure 5.11   Late transient detector for both edges.

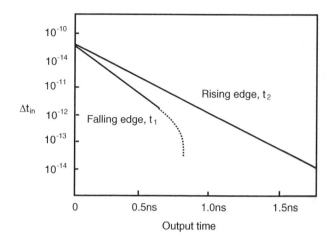

Figure 5.12 Plot of t_1 and t_2.

Both t_1 and t_2 vary with input time Δt_{in}. As the circuit gets closer to the balance point t_1 increases at first, but then reaches a maximum and decreases, whereas t_2 continues to increase without limit.

Typically this leads to an input output plot similar to that shown in Figure 5.12. Since t_2 is always later than t_1, and an early t_1 transition in one direction may be followed by a later t_2 transition in the other direction, the only time of real interest is t_2, which determines the reliability of the synchronizer.

5.4 DELAY-BASED MEASUREMENT

A rather better test method is to keep the clock frequency constant, and low enough to avoid metastability in flip-flop #3, then to vary the delay for which responses are measured [21]. A circuit, which overcomes the two problems of the late transient detector, is shown in Figure 5.13.

Here the output of flip-flop #1 goes to a high threshold (NAND) inverter so that if a metastable level appears at the output, that level will not propagate to flip-flops #2 and #3. Instead, the times at which the output changes are measured. Any change of output value between the time set on the programmable delay, t_{pd}, and the clock period t_c, is detected as a difference between flip-flop #2 and flip-flop #3 at the next clock edge. The clock frequency is constant, and set low enough to ensure that flip-flop #3 is not clocked until the probability of a metastable input to #3 is very low. Metastable events lasting between the time

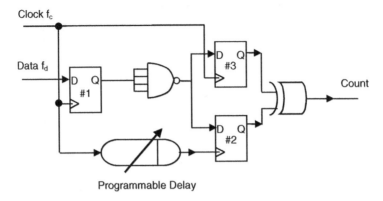

Figure 5.13   Metastability measurement.

set on the programmable delay t_{pd} and the clock period t_c result in a count value N_{pd} being incremented. For each individual setting of the programmable delay, a count of the number of metastable events N is recorded over some long time T. Given that the number of events that last longer than t_c is very small, each count represents the number of times metastability lasted longer than the time set in the delay, minus the high threshold gate time.

The number of events lasting for a time between t_n and t_{n+1} where t_n and t_{n+1} are successive settings of the delay can be found from $N_n - N_{n+1}$, where N_i is the count of events lasting longer than t_i. Only events which actually cause a change in the output of the high threshold inverter can be measured by this method. Because a metastable state in flop-flop #1 must give either a high, or a low output level at the inputs to flip-flops #2 and #3, only approximately half of the trajectories of flip-flop #1 actually change state when the metastability resolves, and can be counted.

A similar method can be used to measure the metastability characteristics of a MUTEX. Figure 5.14 shows how two independent oscillator inputs are used to drive the request inputs, R_1 and R_2. Because they are independent any overlap, positive or negative, can occur between the requests, and the grant outputs are observed a defined time after both requests go high by flip-flops #1 and #2. If neither request has been granted, the MUTEX is in metastability at that time, and after a further delay to ensure that neither #1 nor #2 are metastable the event is counted. A histogram can be constructed as before by incrementing the programmable delay, and subtracting the later count from the earlier, $N_n - N_{n+1}$, thus giving the number of events that resolved between t_n and t_{n+1}.

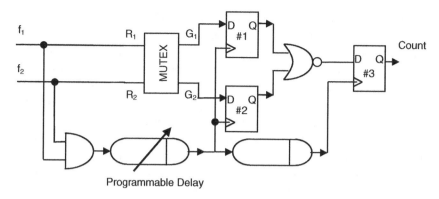

Figure 5.14 MUTEX measurement.

5.5 DEEP METASTABILITY

Unfortunately, events which result in a much longer than normal propagation delay (deep metastability) occur relatively rarely. Such an event will require very much less than 100 ps overlap between data and clock. In the two-oscillator method with oscillator frequencies of 10 and 9.99 MHz the overlap is slowly changing between zero and 100 ns every 1000 cycles of the clock. Times with less than 100 ps overlap will occur in only 1 in 1000 of the clock cycles, so truly metastable events happen much less often than the 100 μs time required for 1000 cycles. Even when they do occur, not all of the resulting output events can be collected if a general-purpose digital oscilloscope is used to collate the data. Because the oscilloscope must store, process and display the histogram, there will be a significant dead time between successive recorded events that limits the actual events recorded, often to less than 1 in 1000 of those generated.

The *MTBF* Equation (2.11) shows that with $f_c = f_d = 10^7$ and $T_w = 100$ ps, an *MTBF* of around 5 minutes requires a synchronization time of 15τ. In an experiment measuring the number of failure events where only 1 in 1000 of these events, occurring every 5 minutes are recorded, more than 1000 times 5 minutes, or 83 hours are needed to measure *MTBF* values of only minutes. Increasing the data and clock frequencies can improve the number of low probability events recorded, but it is not practical to characterize the synchronizer much beyond 16τ.

It is possible to increase the probability of a metastable event by ensuring that the data transition is always within a small time $\Delta t_{in} < 100$ ps from the balance point. If there is a transition within 100 ps on every 100 ns clock cycle time rather than only one in 1000 cycles as with the conventional arrangement there are now 1000 times as many metastable events. Either the same measurements can be made in a much shorter time, or more deep metastable events can be recorded in the same time. In practice a delay-locked loop can be used to hold the data input at a point that gives a 0.5 probability of a high transition in the Q output and a 0.5 probability that it stays low. This point is the balance point that gives many very long output times.

For input times within less than 0.2 ps of the balance point, noise in the sytems and in the flip-flop itself causes the circuit outputs to be nondeterministic and whether the output ends up high or low depends on the noise [18, 19] as well as the input time.

A schematic of a test set-up with an analog DLL is shown in Figure 5.15. Here a 10 MHz clock is passed through two closely matched paths to the data and clock inputs of the device under test. Adjusting the supply voltage to open collector inverters varies the delay in the path to the data input. If the data rising edge is slightly slow when compared with the clock edge, Q will be low on most clocks. If it is fast, Q will be mostly high. A slave flip-flop records whether the device under test has resolved high or low and an analog integrator is used to average the proportion of high to low outputs from the slave. The integrator consists of an operational amplifier with its reference input held at a voltage approximately halfway between the logic high and logic low levels of the slave flip-flop V_{high} and V_{low}. A high slave output causes the integrator output to fall slightly and a low output causes it to rise. The arrangement forms the delay-locked loop in which the voltage supply to the inverters in the data

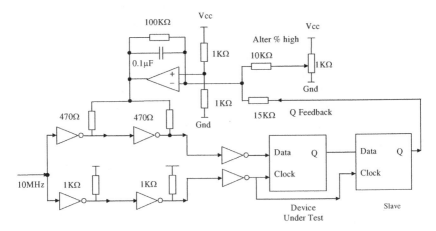

Figure 5.15 DLL controlling data and clock overlap. Reproduced from Figure 3, "Measuring Deep Metastability", by D.J. Kinniment, K Heron, and G Russell which appeared in Proc. ASYNC'06, Grenoble, France, March 2006, pp. 21–1 © 2006 IEEE.

path is increased if it is too slow and reduced if it is too fast. When the device under test is near the balance point its final output is determined by noise and is random. In these circumstances the integrator output will change after each clock period by an amount

$$\pm\frac{V_{high}+V_{low}}{2f_cT_i}$$

where T_i is the integrator time constant. Altering the reference voltage enables the proportion of highs and lows to be set manually. This works by increasing the amount added to the integrator output when the slave output is low and reducing it when it is high. Normally the input voltage is set to $(V_{high} + V_{low})/2$, so that the system settles to a steady state where 50% of flip-flop outputs are high and 50% low, but it is possible to vary the proportions simply by increasing or reducing the voltage. A higher input forces the loop to compensate by reducing the proportion of high slave outputs and vice versa. A separate input to the data delay path supply voltage (not shown) enables a high-speed waveform to vary the delay by around + or − 100 ps, giving the delay distribution with time in ps between clock and data shown in Figure 5.16.

This histogram was obtained using a digital oscilloscope and an approximately sinusoidal input variation source. It shows a 100 ps variation around a balance point of 205 ps where the time axis shown is the difference between the data and clock rising edges.

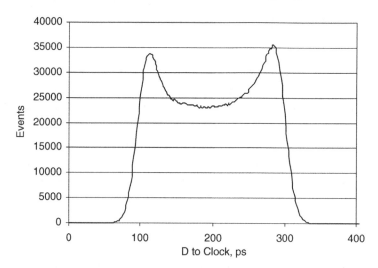

Figure 5.16 D to clock event distribution with time in ps. Reproduced from Figure 4, "Measuring Deep Metastability", by D.J. Kinniment, K Heron, and G Russell which appeared in Proc. ASYNC'06, Grenoble, France, March 2006, pp. 21–1 © 2006 IEEE.

Figure 5.17 Display histogram for clock to Q rising events at 500 ps/division. Reproduced from Figure 5, "Measuring Deep Metastability", by D.J. Kinniment, K Heron, and G Russell which appeared in Proc. ASYNC'06, Grenoble, France, March 2006, pp. 21–1 © 2006 IEEE.

In Figure 5.17 the data collection oscilloscope was triggered from the rising Q outputs and a histogram of the number clock inputs is shown against a time scale of 500 ps per major division. The oscilloscope is in colour grade mode, so that the density of traces at a particular point is represented by the grey level of the pixel at that point on the display. A histogram of the trace density along the horizontal line is also shown in this figure, which represents the number of events passing through the pixels concerned.

In the two-oscillator measurement method, the number of input events per picosecond is constant with respect to D to clock time. In the D to clock histogram of Figure 5.16 the number of input events per picosecond varies with the D to clock time, so for this method it is necessary to correct for the variation. The correction can be done, by noting that for exactly half of the input events on the D to clock histogram the final Q output is high and for half, Q remains low. When D is earlier than the balance point Q is more likely to rise than to remain low and when it is later it is more likely to remain low. Thus if the cumulative number of input events on the D to clock histogram is plotted against time, and the event axis normalized to between −1 and +1, an exact value for the balance point time can be found where the graph crosses zero. In this case the balance point is 205 ps. Because only half the input events cause an output event, the cumulative number of events on the output histogram must be normalized to between 0 and 1. The correspondence between D times and Q times can now be found from the fact that, for a large enough number of events, the number of input events closer to the balance point than the D time must equal the number of output events with an output time longer than the Q time.

Figure 5.18 shows the normalized cumulative number of input events plotted against input time on the left and the normalized cumulative number of

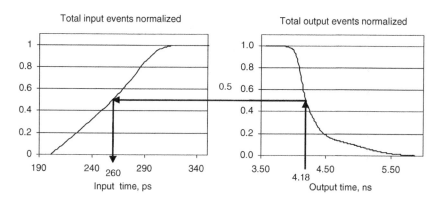

Figure 5.18 Output time to input time. Reproduced from Figure 6, "Measuring Deep Metastability", by D.J. Kinniment, K Heron, and G Russell which appeared in Proc. ASYNC'06, Grenoble, France, March 2006, pp. 21–1 © 2006 IEEE.

output events against output time on the right. In this figure the number of events where the propagation delay is greater than 6 ns is very small, but as the propagation delay falls to near the normal delay of 3.9 ns the number of events rapidly increases to include all of those measured. If we take any proportion of events E where $E < 1$ this proportion can be associated with output times longer than the x-coordinate of E on the right-hand graph, and input times shorter than the x-coordinate of E on the left-hand graph. A small increase in E to $E + dE$ links dE to input times and output times within a very narrow range, and therefore links the input time to the output time. In Figure 5.18 $E = 0.5$ has been identified where below 0.5 all the output events have metastability times lasting longer than 4.18 ns and the input events occur with the D input between the balance point and 260 ps.

The 0.5 point is therefore associated with a unique input time of $\Delta t_{in} = 260 - 205 = 55$ ps and a unique 4.18 ns output time. We can now plot Δt_{in} against output time to give Figure 5.19. This graph shows that inputs less than 55 ps away from the balance point will normally give an output lasting longer than 4.18 ns, so if the synchronization time is set to 4.18 ns failures will occur in a synchronizer with inputs less than 55 ps. It is important to note that the graph of Figure 5.19 is statistical in nature; it is not possible to measure any single input to much better than 1 ps because of the presence of noise. Each point on the graph represents more than one event and is characterized by the number of events that occurred after the output time given by its x-coordinate. The y-coordinate of the point is the input time just greater than the same number of input events. Because input event times are distributed uniformly in this graph over the input event time range, typically 100 ps, random noise has no effect on the input distribution, there are still a constant number of events

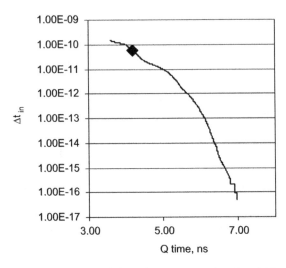

Figure 5.19 Standardized metastability characteristic showing 4.18 ns, 55 ps point. Reproduced from Figure 7, "Measuring Deep Metastability", by D.J. Kinniment, K Heron, and G Russell which appeared in Proc. ASYNC'06, Grenoble, France, March 2006, pp. 21–1 © 2006 IEEE.

per picosecond. The accuracy of the graph depends on the number of events, the more events in the range, the greater the effective accuracy and the smaller the input times that can be plotted.

A significant advantage of this technique is that it allows the results to be presented in an easily understood standardized form, independent of oscillator frequency or number of events. With knowledge of the clock and data frequencies any point on the y-axis, Δt_{in}, can also be converted to give the mean time between failures in a system using Equation (2.8).

If the source of data delay variation is removed, the number of input events close to the balance point will be increased and therefore the ability to measure longer metastable times correspondingly enhanced. Figure 5.20 shows the distribution of input times that results from this change.

Here the deviation from the central value as measured by the oscilloscope is about 12 ps and is very similar to a distribution that would be produced by random noise alone. This is partly because the flip-flop output value, high or low, is at least partly determined by internal thermal noise and partly because there is a significant noise element in the oscilloscope measurements.

The measurement noise can be estimated by producing a histogram of the clock waveform when it is also the source of the oscilloscope trigger. At the actual trigger point voltage the time deviation in the histogram is

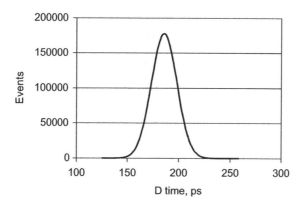

Figure 5.20 Event histogram with only noise variation on the input. Reproduced from Figure 8, "Measuring Deep Metastability", by D.J. Kinniment, K Heron, and G Russell which appeared in Proc. ASYNC'06, Grenoble, France, March 2006, pp. 21–1 © 2006 IEEE.

very low, but at a higher or lower voltage it spreads out to around 9 ps. The specification of the oscilloscope used for this measurement was 9.2 ps.

Because of the relatively large measurement noise component we cannot reliably use Figure 5.20 to assign input times to output times. To overcome this problem the reference input of the integrator in the DLL of Figure 5.15 is changed to produce a range of different proportions of high and low outputs from the device under test. If instead of having a reference input voltage of $(V_{high} + V_{low})/2$, giving 50% high output values we set the voltage to a value of $(3V_{high} + V_{low})/4$, a slave output at V_{high} now gives an integrator input of $(V_{high} + V_{low})/4$ while an output at V_{low} gives an integrator input of $(3V_{high} - 3V_{low})/4$.

The result is that a low output reduces the D delay path three times as much as a high output increases it, so on average, there must be three times as many high outputs as low outputs in order to keep the delay at the balance point. With 75% high values, the centre point of the D distribution moves earlier by exactly the time required to shift 25% of the events from the 0 side of the balance point to the 1 side as shown in Figure 5.21.

We measured the time shifts required to give different probabilities of high outputs and plotted them on a graph showing percent high points against time shift (Figure 5.22). If we assume that the time of input events follows a normal distribution, we can compare this graph with distributions having different values of standard deviation.

The line with the closest fit to the points on Figure 5.22 represents the cumulative probability of a high output for a random input time deviation of 7.6 ps, so we can conclude that the actual distribution has

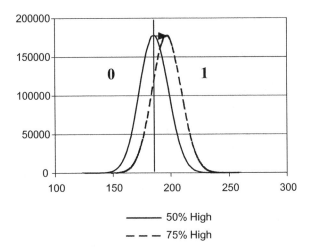

Figure 5.21 Shifting the input distribution. Reproduced from Figure 9, "Measuring Deep Metastability", by D.J. Kinniment, K Heron, and G Russell which appeared in Proc. ASYNC'06, Grenoble, France, March 2006, pp. 21–1 © 2006 IEEE.

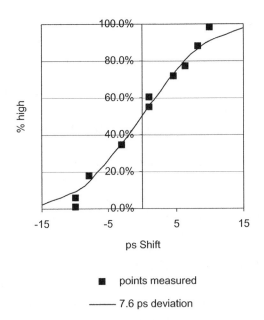

Figure 5.22 Measurement of actual D distribution. Reproduced from Figure 10, "Measuring Deep Metastability", by D.J. Kinniment, K Heron, and G Russell which appeared in Proc. ASYNC'06, Grenoble, France, March 2006, pp. 21–1 © 2006 IEEE.

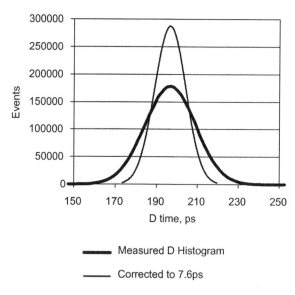

Figure 5.23 Corrected input time distribution. Reproduced from Figure 11, "Measuring Deep Metastability", by D.J. Kinniment, K Heron, and G Russell which appeared in Proc. ASYNC'06, Grenoble, France, March 2006, pp. 21–1 © 2006 IEEE.

a deviation of this value. The corrected input time distribution corresponding to this is shown in Figure 5.23.

For comparison the original histogram measured on the oscilloscope is also shown, demonstrating that measurement noise makes a significant contribution to the raw data. Combining the oscilloscope deviation of 9.2 ps with 7.6 ps as the root sum of squares gives a result of 11.9 ps, very close to that observed.

This method of generating and measuring time distributions at the picosecond level in the presence of noise allows the measurements to be taken another order of magnitude further, as shown in Figure 5.24, by the 7.6 ps line. Figure 5.23 also shows the results of triggering the oscilloscope only on Q outputs that appear after 6.5 ns. We do this by clocking the Q output into two flip-flops, in this case one at 6.5 ns and the other at 50 ns. The oscilloscope is triggered by a delayed clock only if the first flip-flop gives a low and the second a high.

Since digital oscilloscopes collect data continuously, it is possible to measure events occurring well before the actual trigger—in this case 60 ns before. Because only a small number of events last longer than 6.5 ns, the trigger rate is 1000 times slower on average and almost all meaningful events are captured. There are now far more useful events which therefore give greater time accuracy and enable us to go a further three decades down in input time. Unfortunately, we do not know exactly

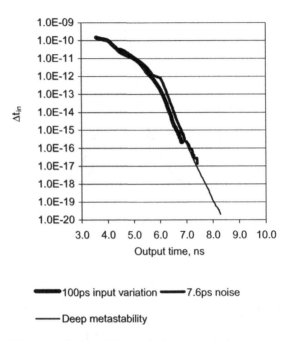

Figure 5.24 100 ps variation, 7.6 ps deviation and deep metastability plots. Reproduced from Figure 12, "Measuring Deep Metastability", by D.J. Kinniment, K Heron, and G Russell which appeared in Proc. ASYNC'06, Grenoble, France, March 2006, pp. 21–1 © 2006 IEEE.

how many of these input events lead to output events over 6.5 ns, but we can count the actual number of triggers to the oscilloscope in a separate fast counter. Normalization of the output histogram avoids the need to know precisely how many of the triggers are converted into oscilloscope traces. Input events causing rising output times longer than 6.5 ns have a very low probability and the probability of any rising clock edge leading to a trigger during the measurement period is given by

$$P = \frac{\text{Output trigger rate}}{\text{Clock rate}} \tag{5.4}$$

Where the output trigger rate is that measured by the fast counter. Around the balance point, the normalized curve of total input events against input time on the left hand side of Figure 5.18 is very linear, that is:

$$K = \frac{d(\text{Total input events})}{d(\text{Input time})} \tag{5.5}$$

Both P and K are constants here. By a similar argument to that presented in Section 5.5, any change in number of output events due to a change in output time is reflected in a corresponding change in input time. In this case the input curve is linear, so it is only necessary to multiply the total output events at any output time on the right-hand side of Figure 5.18 by the constant P/K to get the equivalent input time.

The result of changing the normalized events scale into time is shown in the deep metastability curve of Figure 5.24 that reaches down to almost 10^{-20} s. The graph shows all three measurement techniques, 100 ps deviation, 7.6 ns noise and deep metastability, applied to the Schottky TTL device in Figure 5.24. The experiment was run at 10 MHz for approximately 1000 s in each mode. The resulting range of Δt_{in} is from 10^{-10} s to 10^{-20} s. At the frequencies used in this experiment, 10^{-20} s represents an $MTBF$ of 10^6s (11 days). Other methods as represented by Figure 5.7 are not capable of $MTBF$ measurements of more than a few minutes. It is interesting to note that the value of τ shown by this rather complex bipolar device has at least three different values, 350 ps between 4 and 6 ns, 120 ps between 6.2 and 6.8 ns, then 140 ps beyond 7 ns. This last value cannot be seen using conventional measurement methods. The fairly distinct breakpoint at 6 ns corresponds to an input of about 1ps and is therefore in the deterministic region [18].

The analog techniques described here can easily be replaced by on-chip digital methods. As shown in Figure 5.25, an on-chip measurement circuit is composed of variable delay lines (VDL), the synchronizer under test (SYNC) and control logic. Together they form the feedback loop which adjusts the input time of the synchronizer. There are two variable delay lines (VDL), one is used to vary the delay in the data path and the other is used to vary the delay in the CLK path.

The VDL in the data path is controlled by the 16-bit main counter on the chip. The VDL in the clock path is controlled externally. The main counter replaces the analog integrator in Figure 5.15, and the analog delay lines are replaced by the digital VDLs. In order to vary

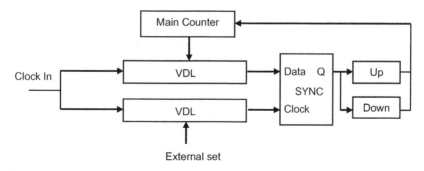

Figure 5.25 Digital deep metastability measurement architecture.

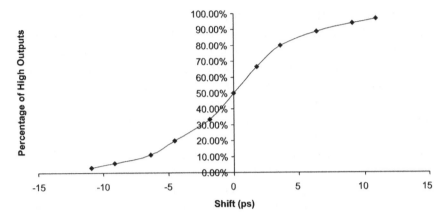

Figure 5.26 Measured input distribution.

the ration between up an down counts, two extra counters are added, which can be preset to divide the number of high ouputs by a preset vaule, and the number of low ouputs by a second number. Because the counters are digital, the ratio can be controlled very precisely, and the input distribution can be measured to an accuracy of around 1 ps, as shown in Figure 5.26

The VDL is based on a current mirror structure which has been proposed by Maymandi-Nejad and Sachdev [22]. As can be seen in Figure 5.27, two current-starved buffer inverters, provide the delay. The current through this two inverter buffer is controlled by a current mirror circuit.

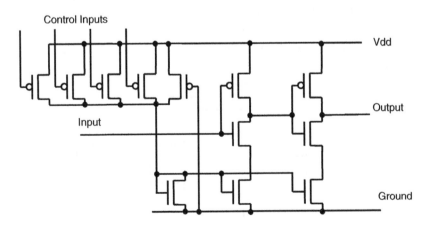

Figure 5.27 Variable delay stage.

Because of the current mirror structure the controlling transistors do not have to be placed below the inverter n-type transistors, so any charge sharing effect is reduced and the delay behaviour of the VDL is monotonic. An appropriate current through the mirror can be adjusted by turning on the four controlling transistors, while one p-type transistor is always on as a base transistor. Here the W/L of the controlling transistors are arranged in a binary fashion. In order to get small incremental delay and large delay range, each VDL includes four cascaded stages, similar to Figure 5.27. The maximum delay of each stage is different and it is possible to achieve an incremental delay of 0.1 ps and a delay range of 0–500 ps. In the digital system deviation of the noise and jitter is smaller than that of the analog design, typically 5 ps being achievable.

5.6 BACK EDGE MEASUREMENT

Many synchronizer circuits have a slightly worse reliability in the second half of the clock cycle than might be expected from projecting the trends measured in the first half of the cycle, and some are significantly worse. While this effect can be estimated theoretically (Equation 2.25), or by simulation, it is not easy to measure using conventional methods.

The techniques described in Section 5.5 can allow measurement of the synchronizer reliability into the region after the back edge of the clock. For the Schottky TTL flip-flop, the low-going clock transition is associated with a 1.5 – 2 ns increase in output time for events after the back edge. This additional delay is clearly observable in the back edge measurements of Figure 5.28 where the input output time characteristics for a long clock high pulse (50 ns) are compared with those for shorter pulses of 4 and 5 ns. The long pulse has little effect on the metastability resolution time, but the short ones add up to 2 ns to metastability resolution times after the back edge.

An estimate of the offset given by Equation (2.25) is shown in Figure 5.29, but it can only be an estimate, since the value of T_d is not known accurately. The reason why the Schottky TTL flip-flop shows such a large increase, is that the difference between the transparent slave propagation time at 150 ps input (less than 3.6 ns) and what it would have been had the slave been metastable (about 5.2 ns) is at least 1.6 ns. The long pulse has little effect on the metastability resolution time, but the short ones add up to 2 ns to metastability resolution times after the back edge. Typically, an inverter delay in 74F technology is

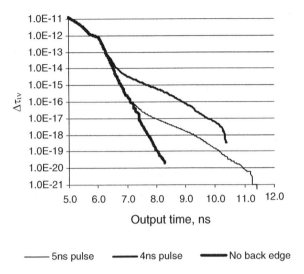

Figure 5.28 Schottky TTL flip-flop input time vs output time for different back edge times.

approximately 3 ns, but this flip-flop has been designed to produce a very advantageous value of τ. The additional delay produced by the back edge of the clock of 1.6 – 2 ns is comparable to an inverter delay, but equivalent to 10 – 15τ and will significantly affect the projected reliability in a one clock period synchronizer.

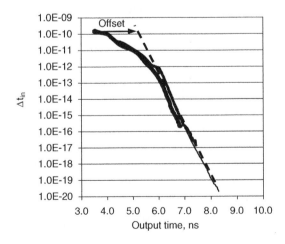

Figure 5.29 Delay change in Schottky TTL flip-flop.

5.7 MEASURE AND SELECT

Process, voltage, and temperature variations in nanometer processes can be an important limit on the performance of systems on silicon. Components such as gates, memories and networks on chip are all affected, but the performance of synchronizers may be degraded by these effects more than other components. Often a system incorporates as many as 1000 synchronizers on the same chip, some of which are critical to the overall performance of the system. The amount of variability in delay in CMOS circuits at different technology nodes is discussed in [24]. At 180 nm, we can expect the variance (σ) to be about 8%. Thus with 1000 circuits one may be 3σ above nominal, or 24% worse than the designed value. At 50 nm, σ is about 15%, so we might expect one synchronizer out of 1000 to have a 45% worse value of τ. There are often only a small number of synchronizers critical to system performance. Current practice is to reduce the effects of process variation by making the transistors in these parts wider than normal so that the deviation is reduced. By making all transistors α times wider, the variation is reduced by $\sqrt{\alpha}$ so a synchronizer four times as big has a variability half of a normal sized one. This technique can use a significant proportion of the system power budget because the current is also increased by α. An alternative is to make a number of smaller synchronizers and select the best. After selection, all the others are powered down, as is the selection circuitry. Power during operation is therefore the same as for a single small synchronizer, but the performance can be improved.

This idea relies on on-chip measurement of failure rates in individual synchronizers, followed by selection.

5.7.1 Failure Measurement

If it is only necessary to compare synchronizers, then accurate measurement is not necessary and a simple comparison scheme is enough.

In Figure 5.30 there are two synchronizers driven by the same clock and data signals. The first is flip-flop #1 and flip-flop #1a, and the second flip-flop #2 and flip-flop #2a. The metastability resolution time allowed in both cases is one clock period When the clock goes low, a third flip-flop, #1b or #2b, is clocked so that the combination of #1 and #1b, or #2 and #2b has a resolution time of only half the clock period. If #1a and #1b resolve to different states, a failure counter (Count 1) is incremented, similarly Count 2 is incremented for #2a and #2b. Since the failure rate

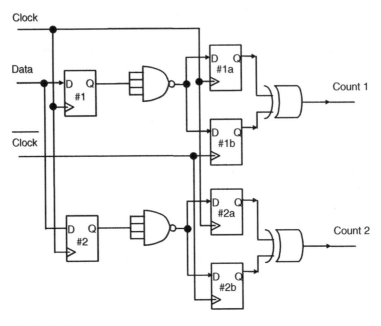

Figure 5.30 Failure rate comparison.

at half the clock period will be much greater than the failure rate at the full clock period the count represents the number metastable events at half the clock period. From Equation (2.10)

$$MTBF = \frac{e^{\frac{t}{\tau}}}{f_d f_c T_w}$$

the number of events counted over a time period T with a resolution time of t will be the square root of the number when a resolution time of $t/2$ is used. Typically if we aim at an $MTBF$ of 3 years (approximately 10^8 s) for a resolution time of t, we can expect a count 10 events in 3 hours at a resolution time of $t/2$. Comparing Count 1 and Count 2 will show directly which is the synchronizer with the best failure rate. Measurement needs to be done on chip, but only needs to be done once in the lifetime of the system.

5.7.2   Synchronizer Selection

On average, selection of the fastest synchronizer from a range of small synchronizers will give a better use of power than increasing the size of

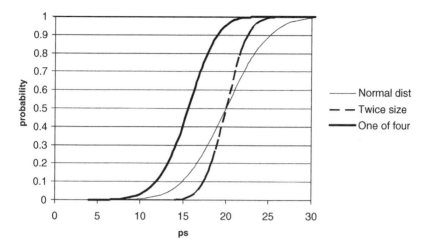

Figure 5.31   Cumulative probability of a synchronizer with a τ longer than a given time.

a single synchronizer. The synchronizer selecting scheme is based on the measurement of τ. Let us assume we have a synchronizer with a τ of 20 ps, and a standard deviation σ of 20%. Assuming that the variability is completely random in the worst case we must allow for a 3.09σ to ensure that fewer than 1 in 1000 of our synchronizers are within specification. This means that the synchronizer time must be set to allow for a τ of 20 + 12.36 ps = 32.36ps. The usual solution to this is to make the width of all transistors in the synchronizer four times larger, so that the deviation is $20\%/\sqrt{4} = 10\%$ Now the worst case is 26.18 ps, but the power is increased by four times.

This is illustrated in Figure 5.31 which shows the probability of a synchronizer having a τ greater than a given time. If the typical synchronizer has a τ of 20 ps, there is a 50% probability of any actual implementation on silicon have a larger value. With standard deviation of 20% the normal distribution curve results, and the curve for twice normal size is also shown. Now suppose we make four standard size synchronizers, measure their τ on chip, and select the best one. After the selection, all the others are powered down, as is the measurement circuitry. Power during operation is therefore the same as for a single small synchronizer, but the performance is improved. The probability of one synchronizer having τ worse than 23.7 ps is 17.8%, but the probability of all four synchronizers having τ worse than this is 0.178^4, or 0.001. Thus we have achieved a worst-case improvement from over 26 ps to under 24 ps. The average value of τ is also improved from 20 ps to about 16 ps by the selection.

These statistics assume that the variability is completely random. This is unlikely to be the case, there will be some correlation between circuits, but note that this correlation will be worse for the transistors in the large synchronizer, because the increase in size is located within a small area, so if the statistics don't work for selection, they won't work for increasing the size.

6

Conclusions Part I

In Chapter 2 models of metastability in a latch are described which enable an estimate of the MTBF of a synchronizer to be made

$$MTBF = \frac{e^{\frac{t}{\tau}}}{f_d f_c T_w} \qquad (2.10)$$

and the shape of a typical histogram of events against synchronization time to be predicted,

$$\frac{\delta t}{\tau}\left[f_d.f_c.T.T_w.e^{\frac{-t}{\tau}} \right] \qquad (2.13)$$

Refinements to the model also allow effects in the deterministic region, and after the back edge of the clock to be estimated.

Chapter 3 describes metastability in simple NAND gate latch circuits, how the nonstandard levels characteristic of metastability can be filtered out, and the design of commonly used synchronizer flip-flops such as the Jamb latch. Versions of the Jamb latch can be designed which show better performance, or are more robust towards supply voltage, process and temperature variation. Circuits useful in asynchronous situations, where metastability may also occur, are the MUTEX and the Q-flop.

In Chapter 4 the noise level in a latch is shown both theoretically and experimentally to be approximately:

$$e_n = \sqrt{\frac{3kT}{C}} \qquad (4.4)$$

Synchronization and Arbitration in Digital Systems   D. Kinniment
© 2007 John Wiley & Sons, Ltd

In a normal system where the probability of all time differences between data and clock signals is constant, noise has little effect on synchronization times, however in the case of a malicious input, it reduces the probability of very long metastability so that adding only a relatively small additional time can return reliability to acceptable levels, in this case:

$$MTBF = \frac{e_n \sqrt{2\pi}.e^{\frac{t}{\tau}}}{f_c V_e}$$

(4.14)

Simulation provides a reliable guide to the nondeterministic part of the MTBF curve, but unless it is augmented by other methods, as in Sections 5.1.2 and 5.1.3 it is insufficiently accurate to predict long term MTBF. The actual reliability of a synchronizer can only be established by measurement of on chip circuits. In Chapter 5 the two-oscillator method provides a simple means to establish the parameters τ and T_w characteristic of synchronizers. Practical measurement schemes include the late transient detector, and the delay-based measurement method.

For measurement in the region of deep metastability where synchronizers normally operate a delay locked loop is necessary to provide sufficient inputs close to the balance point. Difficulties in measuring the distribution of input times in the presence of picosecond level noise in the measuring equipment can be overcome by the technique of measuring the shift in the input distribution as a function of the proportion of high and low outputs. Limitations in the repetition rate of the measuring equipment can reduce the rate of collection of useful output data to as little as 1 data point collected in 1000 generated. This can be overcome by generating only events in the region of interest (deep metastability) so that the repetition rate required is low and normal propagation time events do not obscure the others. Thus a total of seven orders of magnitude improvement over conventional methods of measuring metastability is possible using the methods described.

Measurement of on-chip values for τ can be a valuable tool for optimizing performance or minimizing variability since it is possible to select the best performing circuit from a number, thereby reducing the effects of the variation.

Part II

7

Synchronizers in Systems

The function of a synchronizer is to re-time the data passing from one digital processor (the sender) to another (the receiver), so that it can be correctly interpreted by the receiver. This is only necessary when the sender and receiver are independently timed and even then only when their relative timing is unpredictable. For many systems on silicon it is difficult to provide a common clock at a high frequency with low skew, so independent clocks are used for each processor. Even if the clock frequency is nomininally the same for both sender and receiver in some cases the requirement to re-use processors and IP blocks in new designs or different fabrication processes will often mean the actual clock frequency in any future implementation may not even be known.

7.1   LATENCY AND THROUGHPUT

A very general synchronizing interface is shown in Figure 7.1 which allows for re-timing as the data as well as some simple flow control to ensure that the data rates on both sides of the interface are matched. Each separately timed data item is accompanied by a write data available signal from the sender which indicates that it valid data is present. Since the data available signal originates from the sender it must be synchronized to the receiver clock before it can be recognized. The receiver then transfers the data to an internal register when it is ready, and signals back to the sender that it is safe to send another data item by means of the 'read done' signal. Because the read done signal originates from the time frame of the receiver, it must also be synchronized to the

Synchronization and Arbitration in Digital Systems   D. Kinniment
© 2007 John Wiley & Sons, Ltd

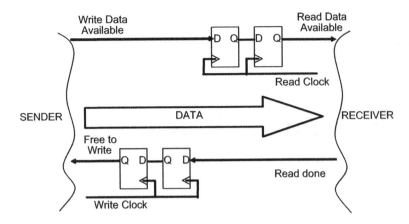

Figure 7.1 Synchronizing interface.

time frame of the sender before it can be understood, and then another item can be sent.

In Figure 7.1 the resolution time of the synchronizer is the delay between the rising edges of the two successive clocks. Normally just one cycle time of the clock is allowed for synchronization. If both sender and receiver are synchronous there will be between 0 and 1 cycle of the read clock between the arrival of data available, and the next read clock then one cycle might be allowed for the synchronizer resolution time. If the time for a single cycle is not sufficient to give a good enough reliability it may be necessary to use two or more cycles for resolution of metastability. With a resolution time of t, a read clock frequency of f_r, and a write clock frequency of f_s there is a delay of between $t.f_r$ and $t.f_r + 1$ cycles of the read clock before the data can be read as a result of synchronization. When the data is read there is a further delay of between $t.f_s$ and $t.f_r + 1$ cycles of the write clock before the next data item can be sent. Often the two clocks have the same frequency, and the time allowed for synchronization is a single cycle. In this case there is a latency of 1–2 cycles, and a minimum turnaround time of 2–4 cycles.

The reliability of the synchronizer depends on the time allowed for metastability to settle. In terms of the metastability time constant, τ of the flip-flop, increasing the delay by a time dt increases the MTBF by a factor of $e^{dt/\tau}$, and in order to keep system failures to less than 1 in 10^8 s (3 years) the total settling time may need to be over 30τ. Since τ is comparable to a gate delay, and high-performance systems may need a clock cycle of less than 10 gate delays, 30τ may be equivalent to several clock cycles. As an example if $\tau = 50\,\mathrm{ps}$ then two clock cycles at 1 GHz are required. Using n clock cycles for synchronization leads to a latency of n to $n + 1$ cycles, and a minimum turnaround time of $2n$ to $2n + 2$ cycles in this simple synchronizer, so that in the worst case the transmission throughput is reduced by a factor of $2n + 2$ when compared with the synchronous case.

There are two ways that a multi-cycle synchronizer can be implemented. One is to divide down the clock frequency by a factor n, and use this reduced clock frequency to drive a simple two-flip-flop synchronizer as in Figure 7.1 so that the resolution time is increased to two cycles, and the other is to pipeline the synchronizer as shown in Figure 7.2. In this scheme any remaining metastability from the first stage of the synchronizer is simply passed on to the second stage for resolution while another sample can be taken for the first stage. The advantage of the pipelined scheme is that read operations can be overlapped, but its disadvantage is that the reliability may be marginally lower as a result of the back edge effect. Because metastable conditions are passed from one stage in the pipeline to the next the $MTBF$ after n stages of the pipeline is proportional to $e^{nt/\tau}$, where t is the resolution time for each stage. $MTBF$ increases exponentially with the number of stages, n, in the same way as the one stage synchronizer would with a resolution time of $n.t$ The difference is that there is an additional edge effect when the clock goes high, and the first latch of each of the pipelined stages goes from transparent to opaque. This effect is usually small, but each extra stage introduces one more back edge delay.

Throughput can be increased by using an intermediate FIFO to separate the read and write operations, but the total time between generation of the data and its availability in the receiver cannot easily be improved. As a result, considerable latency is added to the data transmission between separately clocked zones, and this latency can be a problem, particularly in long paths where a computation cannot proceed until a requested result is received.

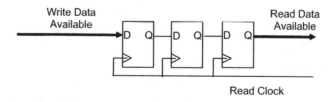

Figure 7.2 Two-cycle pipelined synchronizer.

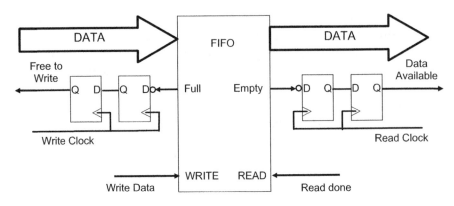

Figure 7.3 Synchronizer with FIFO.

7.2 FIFO SYNCHRONIZER

A synchronizer with a FIFO between the sender and the receiver is shown in Figure 7.3, [25–27]. The effect of the FIFO is to decouple the reading and writing processes so that writing to the FIFO can take place immediately if the FIFO has space available, and reading from the FIFO can be done as soon as the receiver is ready provided there is always some valid data in the FIFO. Because of that the throughput is no longer limited by the round-trip delay of the flow control signals, data available and read done, it can be as high as the FIFO can support. The FIFO itself is normally implemented as a circular set of registers with two address pointers, a read pointer, and a write pointer, as shown in Figure 7.4. The write pointer always points to a free location in the FIFO, and immediately after the data is written in, the pointer is advanced by one. The read pointer always points to valid data that has previously been written, and is also advanced by one immediately after the read operation. To ensure that the writer always points to a free location there must be sufficient space in the FIFO to accommodate any stored data plus the current free location and a further free location to ensure that the write pointer can be moved on without interfering with a read operation. Similarly there must be sufficient stored data to ensure that a read operation can be carried out, and the read pointer moved to new valid data. The write pointer is thus always ahead of the read pointer by at least two locations.

The flags Empty and Full are set by comparing the two pointers. Full is true if there is not sufficient space to accommodate any data items that might be written before the sender can be stopped, and Empty, which is true if there is not enough of unread items in the FIFO that

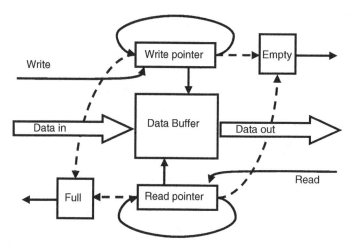

Figure 7.4   FIFO buffer.

might be read before the receiver can be stopped. The read pointer to write pointer distance necessary depends on the delay in stopping the receiver, and the write pointer to read pointer distance depends on the delay needed to stop the sender.

Because the write pointer moves with the write clock and the read pointer moves with the read clock, the Full and Empty flags are not synchronized to either of the two clocks and a synchronizer must be inserted between the Full signal and the sender to ensure it is correctly interpreted. Similarly a synchronizer must be inserted between the Empty signal and the receiver. An n cycle delay in the empty signal synchronizer means that the receiver cannot be stopped for n cycles, and so there must be at least n valid data items in the FIFO. This does not affect the throughput because the delay of n cycles required to synchronize the Empty signal is overlapped with the reading of these n items. Similarly with n spaces the writes can also be overlapped with the synchronization of Full. This scheme does not improve the latency, because the time between writing an item, and reading it is at least n cycles of the read clock plus between 0 and 1 cycle for the Empty signal before further reads can be stopped. Similarly between n and $n + 1$ cycles of the write clock are needed to recognize that the FIFO needs another item. Forward latency (the time between data leaving the sender, and arriving at the receiver) can only be reduced by reducing the time needed to synchronize the Empty signal, and reverse latency (the time between the receiver asking for more data and the data being sent) by reducing the Full synchronization time.

7.3   AVOIDING SYNCHRONIZATION

If the clocks are all phase locked, there is no need for resynchronization of data passing between from one core processor to another, since data originating in one clock domain and passing to the next will always arrive at the same point in the receiving clock cycle. Even if the phase difference is not the same, and may not be known until the clock has run for some time, it is still predictable, and a suitable data exchange time can be found [28, 29].

A simple way to deal with a mesochronous interface, that is, one between clocks derived from a common source, is to use a FIFO [31, 32], as in Figure 7.3. If the FIFO is half full, we ensure that the sender writes one data item into the FIFO every clock period, and the receiver reads one data item every clock period. The distance between pointers cannot now move more than one from the half-full condition, and given sufficient registers in the FIFO the empty and full conditions never need to be tested. No synchronization is necessary.

A simpler interface for mesochronous systems is shown in Figure 7.5. Here only three registers are needed even when the read and write clocks differ by an unknown (but bounded) phase difference. There is a register W for write data, an R register for read data, and an intermediate register X which must be clocked at least $t_{\text{set-up}} + t_{\text{prop}}$ after the write clock, and at least $t_{\text{hold}} + t_{\text{prop}}$ before the read clock where $t_{\text{set-up}}$, t_{prop}, and t_{hold} are the set-up time, propagation time and hold time for the registers.

Putting t_s for $t_{\text{set-up}} + t_{\text{prop}}$ and t_h for $t_{\text{hold}} + t_{\text{prop}}$ Figure 7.6 shows the timing for the register X clock. There are two time intervals where the X clock can be placed. If it is put the first half-cycle of the write clock, new

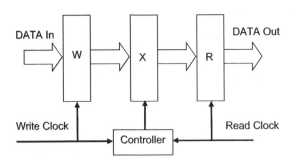

Figure 7.5   Mesochronous interface. Reproduced from Figure 2, "Efficient Self-Timed Interfaces for crossing Clock Domains". by Chakraborty, and M. Greenstreet, which appeared in Proc. ASYNC2003, Vancouver, 12 – 16 May 2003, pp. 78–88., © 2003 IEEE.

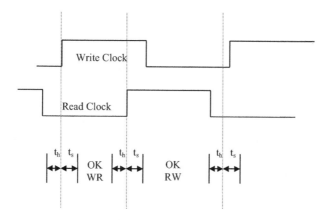

Figure 7.6 Clock timing for register X.

data is written immediately before reading (WR), if it is in the second half cycle data written in the previous cycle is read before writing new data into the X register (RW). Both of these modes allow correct transfer.

The time interval from one rising edge of the write clock to the next is exactly one clock period, and no matter what the phase relationship between the write and read clocks, there will always be a single rising edge of the read clock within the write cycle. In order to allow a correct transfer when writing before reading there must be a time t_s between the write clock rising edge and the X clock. Similarly there must be a time t_s between the X clock and the next read clock rising edge. If the mode used is read before write, there must be a time t_h between the read clock rising edge and the X clock. Similarly there must be a time t_h between the X clock and the next write clock rising edge. The relative size of the two OK intervals will depend on the phase relationship between the write clock and the read clock, The time intervals OKWR and OKRW both vary according to the phase difference between read and write clocks, but the total amount of time for both OK intervals is $t_c - 2(t_h + t_s)$ and provided that $t_c > 2(t_h + t_s)$ at least one of them is always available.

Figure 7.7 shows a circuit which can be used to control the X Clock rise time. It consists of two self resetting latches [33], one set t_s after the write clock goes high, and the other set t_h after the read clock goes high. When both are set, the X clock goes high, thus the X clock occurs t_h after the read clock if the write clock is first, and t_s after the write clock if the read clock is first. Neither latch can be reset until three inverter delays after the last latch was set, so the X clock lasts for three inverter delays.

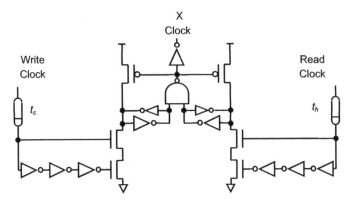

Figure 7.7 X clock controller. Reproduced from Figure 4, "Efficient Self-Timed Interfaces for crossing Clock Domains". by Chakraborty, and M. Greenstreet, which appeared in Proc. ASYNC2003, Vancouver, 12 – 16 May 2003, pp. 78–88., © 2003 IEEE.

The circuit automatically adjusts to changing write–read phase, as can be seen in Figure 7.8. On the left-hand side of Figure 7.8 the write clock is ahead of the read clock, but on the right-hand side the read clock is ahead. The X clock is always generated after both read and write clocks so the circuit is not required to make a decision and consequently there can be no metastability. There is always a latency of approximately one cycle between the input data and the output data.

This scheme can also be used for multiple clock frequencies, provided they are all locked to the same root clock, which is a rational multiple of each. The data transfers can then be done by using a locally generated version of the root clock on each side of the interface.

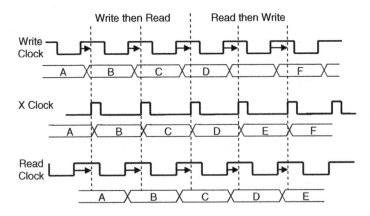

Figure 7.8 Drifting phase. Reproduced from Figure 5, "Efficient Self-Timed Interfaces for crossing Clock Domains". by Chakraborty, and M. Greenstreet, which appeared in Proc. ASYNC2003, Vancouver, 12 – 16 May 2003, pp. 78–88., © 2003 IEEE.

7.4 PREDICTIVE SYNCHRONIZERS

In a plesiochronous system the two clock frequencies on each side of the interface may be nominally the same, but the phase difference can drift over a period of time in an unbounded manner. If the phase difference changes by more than a cycle the simple mesochronous interfaces described above will either repeat data items or lose data items so it is necessary to use some form of flow control. In most cases the two clocks are both driven from very stable crystal oscillators and are to some extent predictable in advance. In this case synchronization can be achieved by predicting when conflicts might occur, and avoiding data transfers when the two clocks could conflict [30, 34, 35].

If the phase change between read and write changes very slowly, it can be detected well before any likely conflict, and this is usually the case with crystal-controlled clock generators. In Figure 7.9 the write clock edge is moving slowly later every cycle when compared with the inverse read clock, which we will use to clock an intermediate X register in Figure 7.10.

When the write clock rising edge gets to within a time t_s ahead of the inverse read clock we detect a possible future conflict between the clocking of the W and X registers. The conflict signal is synchronized to the read clock, and can be used to avoid actual conflict in future by delaying the inverse read clock by $t_s + t_h$. The conflict detection circuit is shown in Figure 7.11 where one MUTEX is used to detect if the write clock is more than t_h later than the read clock, and a second MUTEX detects if the write clock is less than t_s ahead of the read clock.

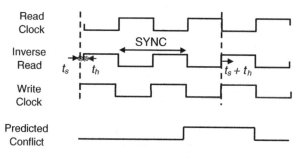

Synchronization problem known in
advance of the read clock.

Figure 7.9 Predicting conflicts.

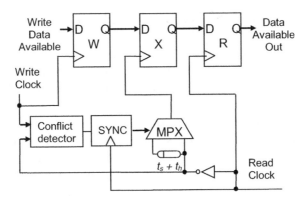

Figure 7.10 Avoiding conflicts.

If both of these conditions are true, there will be a potential conflict. Either of the MUTEXs could produce a metastable output, so the conflict output itself must be synchronized to the read clock to ensure adequate reliability. Figure 7.10 shows how the conflict detection is integrated into a simple interface. The write data available signal is first clocked into the W flip-flop by the write clock. The conflict detector compares the W clock and X clock for potential conflicts, and the conflict signal is synchronized to the read clock. If the X clock and W data are not likely to violate the X flip-flop set-up or hold times it can be used to clock the X flip-flop. If it is likely to be within the conflict window it is delayed by at least $t_s + t_h$, so that the W clock and the X clock cannot conflict. The data available signal is then clocked into the R register when the next read clock occurs. Since there is no metastability, the delay

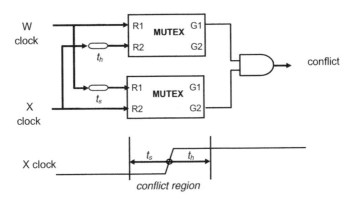

Figure 7.11 Conflict detection.

between the write data available and the read data available is between a half and one and a half cycles, even though there is a two-cycle delay between detecting a conflict and changing the X flip-flop clock timing. The correct operation of this system depends on phase changes over two clock cycles being very small, and it can be used as part of the flow control for data in a FIFO interface giving an average gain of half a cycle in latency over a single cycle synchronizer, and considerably more over a multi-cycle synchronizer.

In a predictive synchronizer of this type there is a trade-off between the sizes of the set-up and hold window, the clock cycle time, and the amount of drift, jitter and noise in the clock cycle that can be tolerated. If the clock drift, jitter and noise is large compared with the cycle time less the set-up and hold time the design of predictive synchronizers becomes more difficult.

7.5   OTHER LOW-LATENCY SYNCHRONIZERS

In practice it may be difficult to achieve accurate and reliable locking between all the clock domains because of jitter and noise in the clock generation circuit, particularly if two different frequencies must be locked together.

Other problems include cross-talk between the data and the clock tree connections introducing noise into both, and dynamic power supply variations between different parts of a clock tree and the data distribution affecting the path delay difference. Temperature changes may also alter the relative phase delay of two clock trees designed in different ways.

All of these effects cause unpredictable variation in the time of arrival of a data item relative to the receiving clock, which is particularly noticeable in high-performance systems using processor cores with large clock trees, and this is likely to increase with smaller geometries [36, 38–41]. Design of interfaces becomes more difficult as these uncertainties increase as a proportion of the receiving clock cycle and it may be simpler to assume that the two domains are independent.

7.5.1   Locally Delayed Latching (LDL)

Schemes like that of Section 7.2 guarantee to deliver synchronized data to a processor core after a latency which depends on the synchronization delay. High-performance systems need at least one clock period for the synchronization time because they have fast clocks and high data rates so, this is an efficient solution, but the majority of applications have lower performance requirements. At low data rates and low clock frequencies the synchronization delay needed to maintain good reliability can be significantly less than the clock cycle time, so synchronization can be combined with some early processing in a single clock period in the first stage of the receiver pipeline. This is done by locally delayed latching [37].

An LDL circuit is shown in Figure 7.12, where the input data is first latched into a register while the controller resolves conflicts between the write request and the read clock. A MUTEX within the controller must decide whether the request occurred before the read clock rising edge, in which case the data is valid, and the next cycle of the read clock can be used for processing, or if it occurred after the read clock, in whch case the data is held in the latch ready for the next clock and valid is false. Conflicts may cause this decision to take some time, and in the worst case the the request wins, but the rising edge of the signal Y1, which clocks the data held in the latch into register R1 is delayed significantly. The time available for processing this data before the next read clock rising edge is then reduced by the controller delay time plus the metastability time of the MUTEX, and any combinatorial logic delays between R1 and R2 must be complete within this reduced synchronization time.

An implementation of a simple LDL input port is shown in Figure 7.13.

In this figure the MUTEX decides between a request and the clock rising edge. If the request wins, the latch is shut to hold the data,

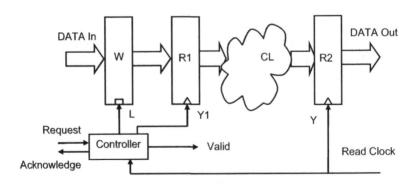

Figure 7.12 LDL circuit.

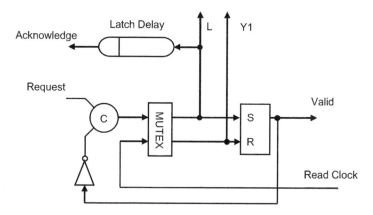

Figure 7.13 Simple input port controller.

and a valid signal sent to the processor. After a delay, to ensure that the input data is captured in the latch, an acknowledgement is sent to the write processor, and the request signal is lowered. The C gate output now falls, and when the clock goes high data is clocked into R1. Valid can now be lowered and acknowledge lowered ready for the next request.

The waveforms are shown in Figure 7.14. Requests can only be accepted when the read clock is low so that they can be processed as soon as it goes high. The first clock rising edge in the figure does not conflict with a request and so can be accepted without any additional dleay. Here L rises after the request latching the data, the request is acknowledged, and L falls when valid is high and request low. When the read clock goes high the data can be clocked into the R1 register.

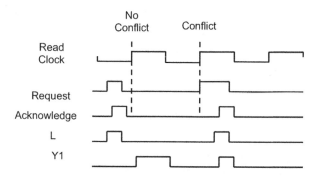

Figure 7.14 Controller waveforms.

The second rising edge conflicts with a request, so that the L signal is delayed by metastability, and consequently the acknowledge is also delayed. Y1 cannot be generated until valid rises after L has gone high, so the rising edge of Y1 occurs later as a result of the resolution of the conflict. This sets a limit on the time allowed for resolution, in this case a little less than half a cycle, and the remainder of the clock cycle is available for the combinational logic to carry out some processing.

The overall latency of this scheme is therefore between 0 and 1 cycle to get the input data into the latch, plus a half-cycle for synchronization. We do not count the second half-cycle because the processor is doing useful work at this time. This compares with between 0 and 1 cycle plus one cycle for a simple synchronizer, where nothing useful can be done until the data is completely synchronized. In a low-performance system where the simple synchronizer usually uses a complete cycle to clock a two-flip-flop synchronizer this can be a worthwhile saving in a system where the clock cycle time is long compared with the combinatorial logic time. Variants of the LDL cntroller can allow for the synchronization plus processing time of the first half-cycle to be allocated in different ways, but the failure rate depends in the end on how much time is allowed for the MUTEX to resolve conflicts. A proportion of the conflicts will always fail because of the bounded time constraint of the clock cycle, and the value of τ for a typical MUTEX is usually worse than that for a well-designed synchronizer flip-flop.

7.5.2   Speculative Synchronization

Very long metastability times occur only rarely, so by assuming that the synchronization time does not normally need to be long, and allowing for recovery from any errors if an event with a long metastability time occurs, the average latency can be considerably reduced. One possible scheme is given below.

7.5.2.1   Synchronization error detection

If the synchronization delay between the read clock and the delayed read clock is reduced the latency is improved, but the probability of failure for each data transmission will be much higher, for example for a 10τ reduction in the resolution time there would be a $22\,000$ times increase in failures, so a 3 year failure rate is reduced to a little more than an hour.

If it were possible to later detect the cases where synchronisation failed, and recover by returning to the system state before failure, the performance of the system could be improved with little penalty in reliability. Note that this does not eliminate the metastability problem, because there will still be cases where metastability lasts longer than the failure detection mechanism allows, and the system state will not be recoverable. Figure 7.15 shows how this can be done by replacing both conventional n-cycle synchronizers with speculative versions in which the data available, or free to write signals are produced early, and allowed to proceed if they subsequently prove to be correct. However, if there is any doubt about their validity at a later time, the R Fail or W Fail flag is raised so that computation can be repeated.

In the speculative synchronizer the first flip-flop must be tested to detect whether it has resolved in a short time or whether it will take longer than average. In order to achieve this we must use a flip-flop which always starts from a known level, and indicates when it has resolved by a monotonic transition. A latch which behaves in this way is the Q-flop described in [43], Figure 7.16, and earlier in Section 3.5. This ensures that half-levels will not appear at the output, and the uncertainty due to metastability appears (in the case of the read synchronizer) as a variable delay time from the read clock to the Q outputs.

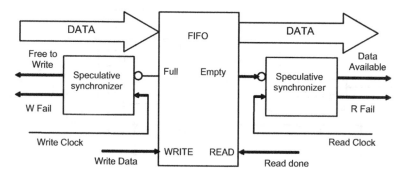

Figure 7.15 Low-latency synchronization.

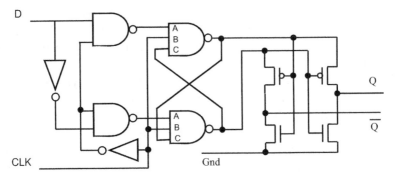

Figure 7.16   Q-flop.

In the Q-flop only clean level transitions occur, because the output is low during metastability, and only goes high when there is a significant voltage difference between Q and $\bar{Q}$.

Once this voltage difference reaches V_1, the transition between a voltage difference of V_1 and V_2 takes a defined time, $\tau \ln V_2/V_1$ so when it is sufficient to produce an output (more than the p-type threshold V_t) the transition to a full output of V_{dd} will take less than 2τ. On the other hand, the time from the clock at which this difference reaches V_1 cannot be determined, and may be unbounded.

The first flip-flop in a speculative synchronizer is made up of a Q-flop latch, and two further latches, #1 and #2 as shown in Figure 7.17. The

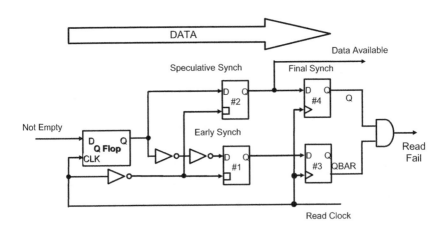

Figure 7.17   Speculative synchronizer.

Q-flop is the master, and the two subsequent latches are slaves clocked by the inverse of the read clock. Because the output of the Q-flop is delayed by two inverters, at the time the clock goes low, latch #1 has slightly earlier version of the Q output than latch #2. Since the Q-flop always starts at a low level, the earlier version will still be low if there is any possibility of latch #2 going metastable.

We will assume that when the read clock next goes high, the outputs of latches #1 and #2 are stable, an assumption that is true to a high degree of probability. At this point they are clocked into flip-flops #3 and #4 which make up the second part of the synchronizer. Because flip-flop #4 must be high if flip-flop #3 is high, there are only three digital possibilities to consider. These three combinations, together with the possibility of metastability when the next read clock occurs are shown in Table 7.1.

In Table 7.1 the first line shows the case where the synchronizer is still metastable after one clock cycle. If the read clock period is designed to give a normal synchronizer probability of being metastable, say e^{-30} failures per clock cycle it will be regarded as low enough to be discounted. If the early and speculative samples in flip-flop #3 and flip-flop #4 are both 0, we can infer that there was no data available signal, and the speculative output was correctly flagged as 0. If the early and speculative samples are both 1, we can infer that there was a data available signal, and the speculative output was correctly flagged as 1. The only other possible case is that the early and speculative samples are 0 and 1 respectively. In this case the output of latch #2 changed some time after the early sample and its value is unreliable. The fail signal is then set and the data transfer must be repeated. The *MTBF* for this arrangement is the same as that of a 30τ synchronizer, because all potentially metastable signals have a full read clock cycle to settle. Even if latch #1 is metastable,

Table 7.1 Synchronizer possibilities.

Early	Speculative	Fail	Comment
0	Metastable	?	Unrecoverable error
0	0	0	No data was available
0	1	1	The speculative output was metastable. return to original state
1	1	0	Normal data transfer

the chances of this happening depend the time allowed for the Q-flop to settle, i.e.

$$\frac{1}{2f_r} - 2t_i$$

where f_r is the read clock frequency, and t_i is the inverter delay. A further

$$\frac{1}{2f_r} + 2t_i$$

is then allowed for the metastability in latch #1 to settle. The chance of it remaining metastable at the end of the read clock cycle is the same as any other synchronizer allowed a total time of

$$\left(\frac{1}{2f_r} - 2t_i\right) + \left(\frac{1}{2f_r} + 2t_i\right)$$

or one cycle time.

This scheme allows a speculative Data Available half way through the read clock cycle, but the probability of a Fail signal is relatively high. The early sample is taken at a time

$$\frac{1}{2f_r} - 2t_i$$

so if the read cycle time is 30τ, and $2t_i = 5\tau$, the time available for the Q-flop to settle is 10τ, perhaps one-third of the cycle. Reliability at this point is reduced by a factor of e^{20}

So instead of a 3 year (10^8 s) $MTBF$ we have an $MTBF$ of about 0.2 s. At a clock frequency of $100\,$MHz this still represents a failure rate of only about 1 in every $20\,000\,000$ clock cycles and the vast majority of read accesses do not fail.

7.5.2.2   Pipelining

If the receiver clock is too fast to allow a single-cycle synchronization time, the read clock can be counted down by two but the fail signal is not generated until the end second cycle, so it is necessary to duplicate the synchronizer so that metastable events created in one cycle are preserved until the second. This is done with the arrangement of Figure 7.18, where a data available is generated within one cycle, but the Fail signal corresponding to that Data Available appears in the following cycle.

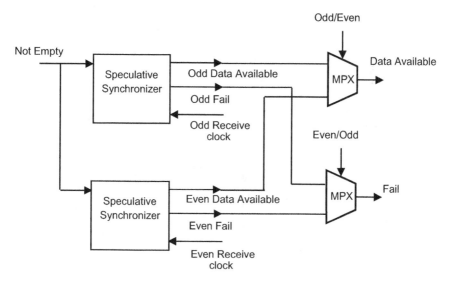

Figure 7.18 Pipelined speculative synchronizer.

7.5.2.3 Recovery

To show how recovery can be achieved after a synchronization failure, we give a simple example in Figure 7.19. Here input data from the sender is normally clocked into the input register, and then added to the R register which has a slave refreshed whenever the input register is overwritten. Read done is signalled back to the synchronizer during the next cycle by the flip-flop B which is clocked by a delayed read clock sufficiently far before the following cycle to enable the read pointer to be updated.

We will assume that the speculative synchronizer uses one read clock cycle for resolution of metastability between The Q-flop and flip-flops #3 and #4. The data available signal appears at the end of the first half of the cycle, and the fail signal does not arrive until after the end of the second half of the read cycle. The receive clock is timed to occur just after midway in the read clock.

If there is a failure of synchronisation such that the speculative data available signal is still metastable at the receive clock time, the input register, the slave, and the flip-flop A may all be corrupted. At the end of the second half of the read clock cycle and when the inverse receive clock occurs, the fail signal will force the flip-flop B to be set to 0, and the R register master will not be altered by the subsequent Receive Clock. Since

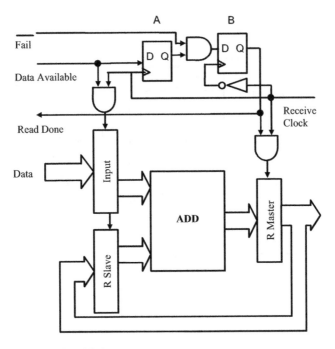

Figure 7.19 Example of failure recovery.

the Read Done is not generated, the read pointer is not updated, and the data available is effectively allowed another cycle to settle before both the input and R register slave are refreshed.

In this scheme only one extra cycle is required every time a synchronization failure occurs, and provided the rate of writing is slower than the rate of reading this extra delay will rapidly be made up. Average latency is now 0–1 read clock cycle for the time between movement of the write pointer which may alter the state of the Empty signal, and the next read clock cycle, plus a half-cycle for synchronization plus 1 in 20 000 000 cycles for possible failures.

This is a relatively simple example, but in general, the strategy is to stop a pipeline when the failure is detected, and recompute or refresh corrupted values. It is important here to ensure that corrupted data, or control states are always recoverable, but this does not usually present a problem. This is similar to schemes in which a fast, but unreliable computation path may need to be backed up by a slower, but reliable one [44]. An alternative recovery scheme involves a shadow register for every data register in the receiving processor [66]. When an error is detected the shadow register is later loaded with the correct value, and a pipeline error is signalled. The processor pipeline can then be flushed and the incorrect data overwritten from the shadow register.

7.6   ASYNCHRONOUS COMMUNICATION MECHANISMS (ACM)

Communication between two asynchronous timing domains usually assumes that every data item sent by the writer to the reader by a channel must be read once and once only by the reader. This is not necessarily the case. There are situations where data may be generated, in the writer, and made available to reader, but not read, for example, a real time clock. The time is a global variable that may be required in one processor in one part of an algorithm, but if its value changes, it is not necessarily needed and the reader cannot be held up waiting for new data. Another situation which differs from the classical interpretation of a channel, is that of a signal from a sensor, where the writer (the sensor) cannot be held up, it must always be allowed to move on to the next reading of the position, temperature, of whatever is being measured even if the data item has not been read. The reader, on the other hand must take note of the value the sensor is producing at a particular point in the control algorithm, and if there is no value it must be held up. A classification of these different communication methods, known as asynchronous communication mechanisms (ACMs) [45–47], is shown in Table 7.2.

In a channel, both the reader and the writer can be held up, and all the previous synchronizers in this chapter are of this type. The pool ACM, allows a real-time clock to write new data items into a shared memory space, overwriting previous values no matter whether they have been read or not, neither the reader nor the writer can be held up. Signal data requires the writer to move on irrespective of the reader situation, but the reader must note new events.

Table 7.2   Classification of ACMs.

	Destructive read (read can be held up)	Nondestructive read (read cannot be held up)
Destructive write (write cannot be held up)	Signal (event data)	Pool (reference data)
Nondestructive write (write can be held up)	Channel (message data)	Constant (configuration data)

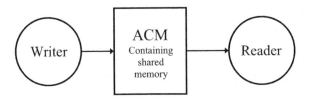

Figure 7.20   Passing data via an ACM with shared memory.

When the writer and reader processes are not synchronized, some kind of intermediate data repository, usually in the form of shared memory, is needed between them to facilitate the data passage. Typically this takes the form of a FIFO as in Section 7.2, but more generally the reader and the writer are separated by an ACM with some internal memory shared by both reader and writer as in Figure 7.20.

The objective of the ACM is to allow reader and writer to be as far as is possible asynchronous with respect to each other while maintaining the requirements of the data to be transferred. Since the writer may be continually writing to the shared memory, locations can be updated just as the reader if accessing the write location. In the FIFO this is avoided by preventing the read and write pointers from ever meeting. This requires some form of flow control in which the writer is not entirely independent of the reader, and the resulting wide pointer separations can lead to long latencies, when there are many more recent items in the memory than the one being accessed by the reader.

Important requirements of an ACM are data *coherence* in which any item of data that is being read from the shared memory must not be changed by the writer (i.e. no writing or reading in part), and data *freshness* where a read access must obtain the data item designated as the most recently written unread item, i.e. the data item made available by the latest completed write access.

The idea of coherence implies that there is mutual exclusion between write and read to the memory. In a software system this is usually provided by explicit critical sections where only one process can enter the critical section at a time. Using critical sections, however, may not be acceptable in real-time systems because the unpredictable waiting time makes it impossible to estimate the precise temporal characteristics of a process. It also makes the writer temporally dependent on the reader and/or vice versa.

This dependence may be in conflict with real-time requirements specified for the reader and/or the writer. For instance, in the Pathfinder mission to Mars, shared memory used for data communications was managed by a complex arrangement of critical sections, priorities and interrupts, but the long critical sections on the shared memory conflicted with the real-time requirements of subsystems, causing occasional system failures.

At this point, it is not necessary to say whether the ACM should be implemented as hardware, or software, and this is one of the strengths of the methodology. The system can be defined in terms of processes and ACMs of different types without committing it to a particular architecture. Only at a later date is it necessary to allocate processes to processors, and communication to hardware or software as the performance requirements of the system dictate.

An ACM is designed to avoid conflicts on data memory without resorting to explicit critical sections, but makes fundamental mode assumptions on the operations of the control variables. Mutual exclusion between writer and reader from the same location is achieved by the state of the shared memory control variables. In other words, the ACM shifts the problem of synchronization from the data memory to the control logic, which typically consists of small shared variables. The general scheme of a hardware ACM is shown in Figure 7.21. In this scheme access to a number of data slots is controlled by the shared variables and the signals wr0, wr1, rd0, rd1, etc. represent actions on those variables. Since read actions rd0, rd1 etc are initiated by the reader and write actions wr0, wr1, etc by the writer sometimes on the same variables at any time, it is possible that two or more actions may conflict. Because of this conflict the flip-flop holding a state variable can become metastable, and any assumption that that the variable always has a digital value does not hold. In practice a hardware ACM is often used by processes where the time between successive write and read operations is very long compared with the metastability resolution time constant, so that

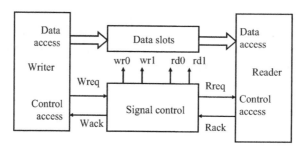

Figure 7.21 ACM with control variables.

the values used are stable enough for each atomic action. The final state of a metastable variable may be true or false, so a robust ACM control algorithm must assume that the variable can settle either way.

7.6.1   Slot Mechanisms

If only one data slot is used for both reading and writing, it will be impossible to make the writer and reader both asynchronous and keep the data coherent because the write and read could occur at the same time to the same register. If two slots are used the system can be made both asynchronous and coherent, but at the expense of freshness, since the system can write to one slot at the same time as reading from the other, but a series of writes one immediately following on the other may lock the reader out of the most recently written data.

Three- and four-slot mechanisms can both preserve coherence and freshness, and here we describe algorithms for pool reference data since these are rather different from the channel message data which can be adequately served by a FIFO.

7.6.2   Three-slot Mechanism

In Simpson's three-slot algorithm [46], the writer algorithm uses three pointers to access three common data items. These are the new pointer, n, the last pointer l, and the read pointer r. Data is written to the location pointed to by n, that is, $d[n]$, data is read from $d[r]$, and l points to the last data item that was written. n is computed from the values of r and l so that it is always different to both of them. If r and l are different, say 1 and 2, n takes up the third value, 3. If r and l are the same, n takes up the next value, 2 if they are both 1, the value 3 if they are both 2, and 1 if they are both 3. To write a data item into the common memory there are three steps. First, the input is written to the next location, then the last pointer updated to point to this location, and finally the next pointer is reassigned to a fresh location different to both read and last. More formally the three steps are:

wr: $d[n] := input$
$w0$: $l := n$
$w1$: $n := \neg(l, r)$

To read, the read pointer is first updated to point to the last item written, and then the item is output as follows:

$r0$: $r := l$
rd: $output := d[r]$

As an example Figure 7.22 shows a new data item (Item 4) being written.

In (a) $l = 3$, $n = 1$, and $r = 3$. This means that the last, and freshest item was written into slot number 3, and the input (Item 4) must be written into slot number 1, $d[1]$. Following the write, (b), l updated to 1 and n updated to 2. In this example we are going to assume that the reader makes an access as soon as item 4 is written. In that case, (c), r becomes 1, following the read and n is changed to 2. From this it can be seen that there is always a third slot for the next item, so that read and

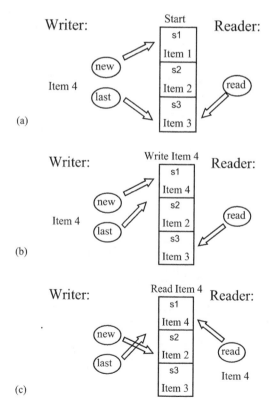

Figure 7.22 Writing a new item in a three-slot ACM.

write can never clash (coherency), and the read pointer always points to the latest fully written item (freshness).

7.6.3 Four-slot Mechanism

The algorithm of the four-slot mechanism is as follows

wr: $d[n, \neg s[n]] := input$
w0: $s[n] := \neg s[n]$
w1: $l := n \| n := \neg r$

r0: $r := l$
r1: $v := s$
rd: $output := d[r, v[r]]$

Here the writer and reader processes are single-thread loops with three statements each. The mechanism maintains the storage of four data slots $d[0, 0]$ to $d[1, 1]$ and the control variables $n, l, r, s[0...1]$, and $v[0...1]$. Variables $n, l,$ and r are single bits indicating next, last and read location for the upper or the lower pair of slots, as shown in Figure 7.23. Individual slots can be addressed by $s[0...1]$, and $v[0...1]$ which are vectors of two single bits. The statements *wr* and *rd* are the data accesses and the other statements are used by the writer and reader to choose slots and indicate such choices to the other side. The statement *w1* is a parallel assignment.

In Figure 7.23 $n = 0$ points to the upper pair of slots, and $l = 1, r = 1$ point to the lower pair. $s[0] = 1$ and $v[0] = 1$. There are four items

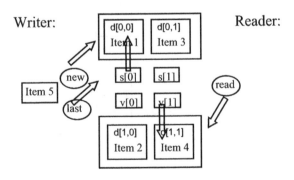

Figure 7.23   Four-slot ACM.

already written into the slots, of which Item 4 is the freshest, and a new item (Item 5) is about to be written. The statement

$$wr: d[n, \neg s[n]] := input$$

writes it into the upper left hand slot, $d[0, 0]$. Statement

$$w0: s[n] := \neg s[n]$$

changes $s[0]$ from 1 to 0, and

$$w1: l := n \| n := \neg r$$

makes $l = 0$, and $n = 0$, thus the next item to be written would go into $d[0,1]$. A read updates r to 0, and $v[0]$ to 0 so that the next read comes from $d[0,0]$,which is the freshest item. Successive writes without a read intervening will go to alternating slots in the same pair (upper or lower) until the value of n changes, and this does not happen unless r changes. Successive reads without an intervening write cannot alter l, n or s, so are always read from the same location. Because $n := \neg r$ after a write the next write is always to a different pair from the next read (coherency) and because the read is always from the same location as the last complete write, we have freshness.

The most important properties of these ACMs are potential full asynchronism, data coherence, and data freshness. The algorithms do not specify the relative timing relationship between the reader and writer processes. This implies that they may be allowed to operate entirely independently in time. The control variable l in both cases is used to steer the reader to the slot which the latest wr statement visited.

The one relative timing specification implied in both algorithms is that within the reader and writer processes, the statements must be executed in the order specified and any statement may not start without the previous one having completed, that is, they are atomic. The four-slot ACM has been shown to maintain full data coherence and data freshness when the ACM is operating in fully asynchronous mode and when the metastability assumptions about the writer and reader statements hold. This is true even when the individual statements are assumed to be globally nonatomic, i.e. including such extreme cases as any writer statement taking as long as an infinite number of reader cycles to complete. The three-slot ACM has been shown to maintain full data coherence and data freshness, but only if statement $r0$ is assumed to be atomic in relation to statements $w0$ and $w1$.

7.6.4 Hardware Design and Metastability

The four-slot ACM design provides maximum safety by adhering to the sequential specifications of the algorithm. However, one of the main arguments for a hardware solution compared with a software implementation is the potential of maximizing the speed of the ACM. This is the aim of the three-slot ACM implementation in Figure 7.24. Since it is known that the three-slot algorithm provides data coherence and data freshness if $r0$ is regarded as atomic relative to $w0$ and $w1$, we must use an arbiter to contain metastability, and this has the side effect of providing this atomicity. Each of register blocks in this figure has correspondence with one of the variables, so we can deduce that since the hardware solution of the three-slot ACM has similar form to the algorithm itself, and will have the correct behaviour.

In the three-slot algorithm, $w1$ changes n, $w0$ changes l, and $r0$ changes r. We must ensure that there is mutual exclusion between statement $r0$ and the statement sequence $w0 + w1$. This is done by a MUTEX which grants access to the registers either to the read control or to the write control, but not to both at the same time. Any metastability is contained within this MUTEX circuit. The block which implements the statement $w1$: $n := \neg(l, r)$ ensures that the control variable n is assigned a value different from the current values of both l and r. In practice, this can be done in hardware by a look up table implementing the matrix $\neg(l, r) = ((2,3,2), (3,3,1), (2,1,1))$.

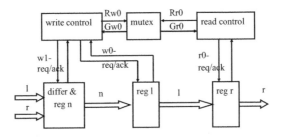

Figure 7.24 Self-timed design outline for a three-slot ACM.

7.7   SOME COMMON SYNCHRONIZER DESIGN ISSUES

Synchronizers are often omitted where they are needed in a system, and sometimes new synchronization circuits are invented which seem to eliminate, or avoid the problem of metastability [48]. These problems are common, but it is not easy to check a design for metastability. First it is necessary to find the conditions that might cause a circuit to go metastable, and then to demonstrate that it actually can be metastable. Metastability does not show in digital simulation, and might not even show up in simulation at the SPICE level because it depends on very small time differences, so special tools are often used to check for asynchrony. Unfortunately they are not perfect. Often they depend on the tool recognizing where clocks are independently timed, information which may not be available in the circuit description without special user input. Any signal whose time reference originates in one processor, and is to be used in another time frame is a potential cause of metastability and usually needs synchronization. Other causes of problems are circuits specially designed to detect metastability, take corrective action, and thus avoid the consequence of long time delays and/or unreliability. These must always be regarded with suspicion. Detecting metastability may itself require a decision, is the circuit metastable or not? And if it does, then the time to make that decision may also be unbounded.

Synchronizers made up of many parallel flip-flops have also been proposed [49]. Some designs can give an advantage, others may not, but generally the advantage is small, and the power required to support a multiple flip-flop synchronizer might be better spent in improving the gain bandwidth product of the flip-flop itself.

7.7.1   Unsynchronized Paths

All designs should be carefully screened to check for signals passing between independently clocked domains. In a safe system all of these should be synchronized.

7.7.1.1   No acknowledge

A common source of an unsynchronized signal occurs in a data transfer where the availability of data is signalled by a request from the sender accompanied by data. Most designers recognize that the request has to be synchronized to the receive clock before the data can be admitted, but sometimes it is forgotten that any acknowledge signal sent back from the receiver to the sender must also be synchronized. This is shown in Figure 7.25, where the acknowledge is asynchronous as far as the sender is concerned and must be synchronized to the sender clock.

7.7.1.2 Unsynchronized reset back edge

Another problem can be caused by an unsynchronized global reset. Why should this be a problem? The purpose of the global reset is to set all flip-flops in a system to a defined state, and the clear or preset inputs used to do this override any clock inputs so that the system will be put into the correct state irrespective of any clock activity. But when the reset signal finishes, however, it may be that the D input to a flip-flop resulting from some previous stage is high, where the low reset has forced the Q output low. Now as the reset trailing edge can occur at any time, when it returns high metastability may occur because the clock happens at about the same time.

Thus, the leading edge of the reset does not need to be synchronized, but the trailing edge does. A trailing edge synchronizer is shown in Figure 7.26 where the clear cannot return high until after the clock edge.

These are just two examples of unsynchronized sneak paths which may cause problems and are not easy to detect.

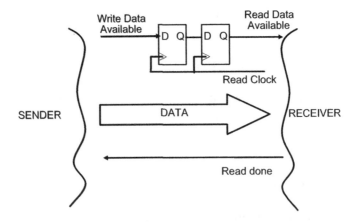

Figure 7.25 Unsynchronized acknowledge.

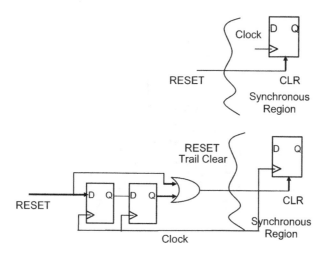

Figure 7.26 Asynchronous reset, and synchronized trailing edge reset.

7.7.2 Moving Metastability Out of Sight

Innovative designers, convinced that metastability is an avoidable problem in their design, have in the past produced ideas to eliminate it, or at least to avoid its consequences. These are not usually safe, and just end up moving the place where metastability occurs.

7.7.2.1 Disturbing a metastable latch

At the circuit level, it is clear that a metastable state is very sensitive to disturbance, and so a small externally applied stimulus could easily knock a metastable flip-flop out of its state and cause it quickly to resolve to a stable high or low. Typical of such ideas is Figure 7.27 which is an attempt to speed up metastability resolution.

How it is supposed to work The latch in this Figure consists of two cross-coupled inverters, a clocked data input and a reset. An additional relatively small transistor has been put in to disturb the latch and help it resolve to predefined state if it ever becomes metastable. There are two possible variations of this scheme. First, the idea may be to apply the pulse unconditionally whenever the latch is clocked. In this case, if the data changes just as the clock goes low the latch is left in a metastable state. The additional *p*-type transistor is

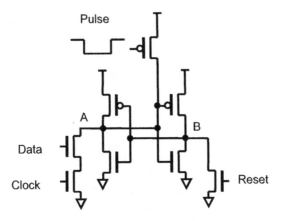

Figure 7.27 Disturbing a metastable circuit.

always pulsed on at this time so that node A is pulled up by a small amount at first, and then rapidly moves high, under the influence of the feedback loop.

Why it doesn't work   The problem here is that without the extra transistor the latch is perfectly symmetrical, so that metastability occurs when the data goes high just a little before the clock goes low. With the extra pulsed transistor added node A is pulled up when the pulse occurs, but if the data change was a little earlier, node A would already have started to move down, and could actually be pulled back up to metastability by the pulse. The data can always change at a time which gets the circuit into metastability with any reasonable pull-up transistor. Only if the extra transistor is very large will node A always get pulled up quickly. But if it always gets pulled up, node A always ends up high. That could be achieved by connecting it straight to V_{dd}, and no decision is being made.

7.7.2.2   The second chance

The second scheme is to arrange the pulse to only come on when the circuit is metastable, so that fast decisions are left alone, and only the slow ones are disturbed.

How it is supposed to work   Metastability is first detected, and then resolved by the pulse, so the circuit always works in a limited amount of time.

Why it doesn't work   Again, the extra pulsed transistor causes node A to be pulled up when it is at a metastable level. Similarly if the data change was a little earlier, node A might have started to move down, and could be pulled back up to metastability by the pulse. Exactly where this new metastable point is depends on the circuit used to detect metastability. It is possible to imagine a metastability detector which gives a half output when node A is almost at the metastable level, and so half turns on the extra transistor. No matter how the circuit is designed, it is always possible to find input conditions halfway between those that give a final high outcome on node A and those that give a final low outcome.

One form of this circuit that does give an advantage is described in Section 3.7. It has two extra transistors, one on each node. Metastability still happens, but the circuit has an improved τ, not because metastability is disturbed, but because the loop mutual conductance is increased when the circuit is metastable by the extra current in the latch transistors

7.7.2.3   Metastability blocker

If metastability cannot be prevented from occurring, and cannot be eliminated in bounded time, perhaps its effects can be blocked from the following system. The circuit of Figure 7.28 is supposed to prevent metastability from affecting the synchronized input of a processor core.

How it is supposed to work   First Reset clears both the latch and the synchronizing flip-flop. When the clock goes high, the input is selected by the multiplexer, and the SR latch is set. When the clock goes low, the latched input is selected by the multiplexer, and a clean signal clocked into the flip-flop next time the clock goes high.

Why it doesn't work   The metastability has been moved from the flip-flop to the latch. Input and clock are still asynchronous, and if the clock

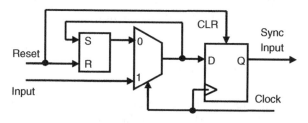

Figure 7.28   Metastability blocker.

goes low at around the time the input goes high, a short pulse could put the latch into metastability. Now there is a half cycle delay before the latch metastable level is clocked into the flip-flop, so the reliability is worse than the normal synchronizer. In this design, there is an even worse problem. It is possible for the input to go high just before the clock goes high. Now there is a short pulse on the D input to the flip-flop, which may or may cause metastability in that as well.

7.7.3   Multiple Synchronizer Flops

Some synchronizer ideas are based on using many synchronizers in parallel to get some speed advantage, or eliminating the time overheads of synchronizing the control flow signals.

7.7.3.1   The data synchronizer

Synchronizers are usually placed in the data available and acknowledge paths. The data available signal cannot go high until the data is stable, and following synchronization, the synchronized data available signal is used to clock the data into an input register. Why not save time by using two flip-flops in every data bit path?

How it is supposed to work   The read clock in Figure 7.29 samples every data bit when the clock goes high. There is then a one-clock cycle delay to resolve possible metastability between the first and second flip-flop in the data path. This second flip-flop then acts as the input register, so extra time to clock the input register is avoided, and latency improved.

Why it doesn't work   Small differences in delay between each of the data bits will mean that some bits get correctly set, and others don't if the clock edge occurs at around the time all the data bits change. The data clock into the input register is then incoherent, some bits being left at their previous value and some set to the new value.

7.7.3.2   The redundant synchronizer

Another interesting idea is the redundant synchronizer, in which many flip-flops are clocked, and the probability of some combination of all

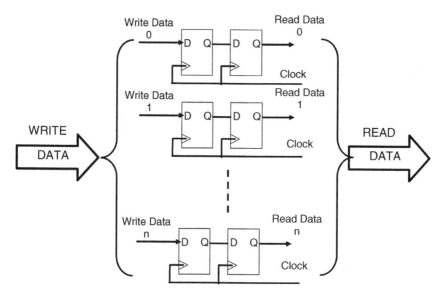

Figure 7.29 Data synchronizer.

the flip-flop outputs remaining metastable after a time t can be made vanishingly small if the number of flip-flops is large enough [49].

How it is supposed to work In Figure 7.30 the input data available is clocked into many flip-flops simultaneously, and then a combinational

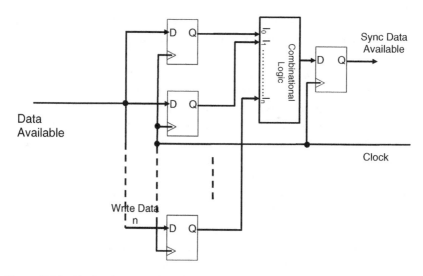

Figure 7.30 Redundant synchronizer.

logic block is used to compute some simple function of all their outputs. When the clock and data coincide, each of the flip-flops may resolve to a high, or a low level, or may remain metastable with a probability of $P(M) = e^{-t/\tau}$ which decreases with time. Assuming that the data and clock edges of all the flip-flops are close to the balance point, the probability of the ith flip-flop resolving to a high after a time t is:

$$P^i(1) = \frac{1 - P^i(M)}{2}$$

The probability of is resolving to a low is:

$$P^i(0) = \frac{1 - P^i(M)}{2}$$

and the probability of it being metastable is: $P_i(M)$

Now let us suppose that the function in the combinational logic block is AND, which appears to be the best choice.

Using p_1 for the average probability of a flip-flop resolving to a high, and p_M for the probability of it still being metastable, it is argued that with n flip-flops the probability of the output of an n-input AND gate still being metastable is:

$$(p_1 + p_M)^n - p_1^n$$

As n increases this is a monotonically decreasing function, so if n is big enough it will eventually become very small. This means we can reduce the probability of metastability to an arbitrarily low level.

Why it doesn't work   It is indeed true that the probability of metastability is much less as n increases if we only look at the case where the data available change is exactly at the balance point where the probability of flip-flop low outcome is equal to the probability of a high outcome. In a real system the probability of a data input changing at any specific time is just the same as the probability of it changing at exactly the balance point so we must consider, what happens if a data available input happens slightly earlier than the balance point. Now the probability of a high flip-flop is increased, and though the probability of metastability is reduced, it is only reduced by a small amount.

Sections 4.2 and 5.2 show that typically the distribution of metastable events with time shift is normal, and the probability of a high out put will increase as the data time becomes earlier according to the cumulative distribution. Using this fact we can compare a synchronizer with one flip-flop and a one-input AND function, which is, of course the normal synchronizer with a two-flip-flop solution with a two-input AND gate.

Putting $n = 1$ we get the probability of metastability to be p_M as expected, this is the normal synchronizer reliability, but putting $n = 2$ we get $2p_1 p_M + p_M^2$. Thus moving the input a few picoseconds earlier we can increase the probability of high outputs from the two flip-flops at the same time by quite a lot as well as reducing the probability of metastability by a little. Suppose the input is earlier by about 0.8σ, where σ is the standard deviation of the timing jitter distribution (normally about 5 ps). This increases p_1 from 0.5 to about 0.79, and reduces p_M by a factor of 1.377. The net result is that probability of metastability, $2p_1 p_M + p_M^2$, is a factor of 1.674 higher than before, so two flip-flops are worse than one. Maybe more might be better.

By shifting the data early enough we can increased the probability of high outcomes to such an extent that the most likely pattern in an n-flip-flop synchronizer is $n - 1$ highs, and one metastable output, leading to a high probability of a metastable AND gate output. The probability of metastability in any one flip-flop is significantly reduced, but the probability of metastability in the AND output is approximately $n.p_1^{n-1} p_M$. Here there are n cases where one input is metastable, and the other $n - 1$ other inputs all need to be high. In order to get probability of metstability better than the one-flip-flop case we need to move the input by 1.39σ, so that p_M drops to 0.5 of its original value, and p_1 rises to 0.917. This cannot be achieved unless $n > 31$, so the AND gate needs 32 inputs. Beyond 32 there is a theoretical gain in reliability over the simple synchronizer. On the other hand there is an extra delay in the 32 input and gate which effectively reduces the synchronizer resolving time, and the strong possibility that a flip-flop that becomes metastable with an input offset greater than the standard deviation is likely to have a fabrication defect, and therefore to be slow.

8

Networks and Interconnects

Networks of independently timed processors are very common because of the difficulty of linking the timing of two or more processors each of which is physically distinct from the others. The internet is an example of a network of processors in which vast amounts of data move between many millions of personal computers and servers, each of which operates with its own independent clock. It is impractical to link all the clocks in all the PCs in the world, so the data must be re-synchronized as it arrives at the input to each PC. This is not really a problem, as the data rates are relatively modest, usually less than 10 MHz, and any delays resulting from synchronization are small compared with the overall communication link delay. The development of integrated circuit fabrication technology has also produced networks on a much smaller scale on chip, where it is also difficult to distribute a single clock. Increasing chip size and reducing metal interconnect widths mean that the delays produced by a connection can be much greater than the delays produced by a gate, or even a processor. Not only are they large, but they are also difficult to predict at the design stage, so the design of a completely synchronous system on chip in which all the registers are controlled by the same clock becomes much more difficult.

8.1 COMMUNICATION ON CHIP

A system on silicon will normally consist of many processor modules connected by a system of communication links. The processors themselves may come from different design sources, have different capabilities,

Synchronization and Arbitration in Digital Systems D. Kinniment
© 2007 John Wiley & Sons, Ltd

and will almost certainly have different clocking schemes, or even be internally completely asynchronous. To allow the design costs of the system to be amortized over a number of applications, some of which may be unknown at the design time, the connection fabric must be capable of adapting to data flows of differing bandwidth and latency.

In traditional designs the signal paths are arranged in an *ad hoc* manner, in which dedicated links carry data from the output of one processor to the input of another as necessary. In dedicated links, the signal paths are not shared, each route from one processor to another has its own physical link, and when data is not being transmitted between processors, the link is idle.

As the size of the system increases, more links must be put in and the number of links used increases at a rate which depends partly on the performance required, and partly on the characteristics of the system itself. Figure 8.1 shows the typical layout of a system on chip which follows this approach, and illustrates how the density of wiring increases in the centre of the chip as the number of processor modules increases. The upredictability of the layout is sometimes characterised as 'spaghetti'. Typically, the number of links increases asymptotically in an order between $O(n)$ and $O(n^2)$, where n is the number of processor modules being connected. A relatively low-performance architecture

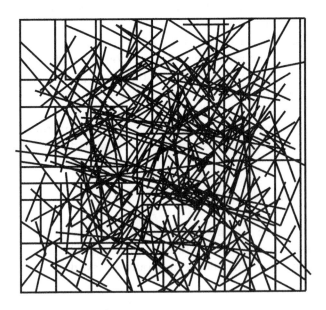

Figure 8.1   Point-to-point links.

might have a relationship between processors and links in the order $O(n)$ and a high-performance one $O(n^2)$. At the same time the average length of the connections increases $O(\sqrt{n})$ because the dimensions of the chip increase at that rate. Relationships like these are the basis for the so-called 'Rent's Rule' [50,51], a measure of the number of external terminals required for an n-sized sytem of computing elements. According to empirical evidence pins increase according to $Pins = Kn^p$ which is $O(n^p)$, where K is the number of pins on each processor block, and p is the Rent exponent, normally in the range $0.5 - 0.75$. This has been shown to hold both empirically and theoretically for a large number of real systems.

The rate of increase of links in a system with a rent exponent of p is normally $O(2p)$, and so the total wire area for a low-performance architecture increases at number of wires times length of wires or $O(n\sqrt{n})$, and for a high-performance system $O(n^2\sqrt{n})$. High end architectures do not scale well with these metrics, and the build-up in wire density in some places shows that the approach is not practical for large systems. Efforts to deal with the scalability and design management problem have led to the use of common busses for data.

The bus-based approach is shown in Figure 8.2 where a number of busses are used to transmit data across the chip. Each bus can carry only one data item at a time so the total number of parallel items is the same as the number of busses. This approach is much more regular than the point-to-point scheme, so layout and design management are easier, but performance can be restricted by the available bus bandwidth. Busses also introduce the problem of bus arbitration. If two or more

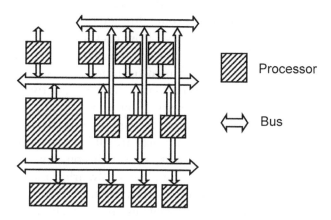

Figure 8.2 Bus links.

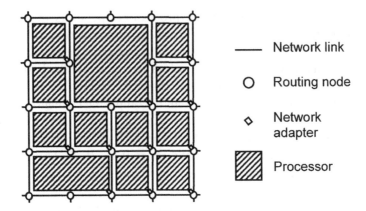

Figure 8.3   Network on chip.

processing modules request the use of the bus, only one request can be granted at time. An arbiter must be used to control the bus access, and that arbiter will be subject to metastability if both requesting processors have independent timing. Issues such as how many busses are needed to achieve the required performance, and how to allocate processors to busses can also make the bus-based approach difficult to scale in large systems.

A more regular and scalable method is to use a network on chip [52,53] as in Figure 8.3.

The network consists of a number of links, arranged in a regular manner around each processor. The processors are connected to the network by a network adaptor which contains a synchronizer so that data from and to the network can be synchronized to the processor read clock. Data is normally organized into packets rather than individual data items, so that information on the preferred route through the network, number of bytes in the packet, and the priority of the packet can be encoded into a header. Routing nodes control the path of each packet at the junction of links so that a packet can be routed from one processor to another. Within a network it is possible for a routing node to have two packets on its inputs, both competing for priority. If the node output ports are free, arbitration must take place between the inputs to decide which packet should be taken first. If the output ports are all busy the input packets are all held up until one or more output ports become available, and if more than one becomes available simultaneously, arbitration may again be necessary to decide which route to allow first.

Performance and congestion problems are not eliminated in the network approach, but system simulation can be used to estimate the required bandwidth of each physical link, and provided sufficient link wiring is used to build adequate bandwidth into the links with the most traffic, software-based routing algorithms can make adjustments to the paths used by the network later.

The flexibility of this approach can replace busses and dedicated signal wiring by providing enhanced performance, scalability, modularity, and design productivity

8.1.1 Comparison of Network Architectures

The important parameters of a network on chip (NoC), are the effective bandwidth of the network, the power required to support the bandwidth, and the wire area of all the links. A proper investigation of the most appropriate NoC architecture for a particular application would need a detailed study of a wide range of alternatives, but it is possible to make very simple comparisons for a limited number of NoC architectures [54]. In order make these comparisons several simplifying assumptions must be made, and here we will assume that:

1. The bandwidth of a link is inversely proportional to its length squared. This reasons for this are discussed fully in Section 8.2, but the implications of this assumption are that in a chip with n processing modules, a link whose length is proportional to the chip side dimensions has a bandwidth $O(1/n)$.
2. The total network bandwidth is proportional to the number of links times their average bandwidth.
3. Each link is fully utilized, so that the power consumption for a link is proportional to the capacitance of all the wires times the link bandwidth, and therefore the length times the bandwidth if the number of wires is constant.
4. The area occupied by a link is proportional to its length, if the number of wires is a constant, and so the total wire area of a network is the number of links times the average length.

A very simple NoC architecture is shown in Figure 8.4, in which every processing module is linked to every other module. The number of links required to do this is $n(n-1)/2$ which is $O(n^2)$. As the array of n processors gets bigger the pattern of links remains the same, even though the number of links increases, so the average link length increases $O(\sqrt{n})$. Using the assumptions above, the total bandwidth is number of

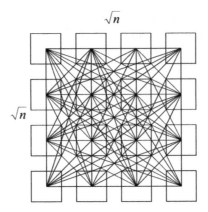

Figure 8.4   Point-to-point architecture, all points to all points.

links times average bandwidth, or $O(n^2.1/n) = O(n)$. The wire area is $O(n^2\sqrt{n})$, and the total power dissipation is $O(n\sqrt{n})$.

In a fully connected point-to-point network it is rarely possible to have all the links fully utilized, but if it were possible it would also give a very good utilization of the processing nodes and lead to a high-performance system. In a point-to-point network the number of terminals needed to add an external processor to an n processor system is n, so the Rent exponent is the maximum possible of 1.

Using a single bus to connect all the processing modules as in Figure 8.5 leads to a considerable reduction in the number of links,

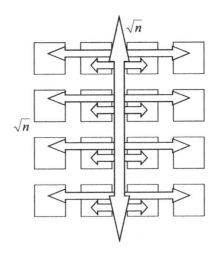

Figure 8.5   Single bus architecture.

as only one is needed, that is, the bus itself. With only one bus all the communication bandwidth required must be supplied by this single bus. To make a comparison with the point-to-point architecture, the bandwidth capability of both should be the same, and since the length of the bus is $O(\sqrt{n})$, there must be n^2 parallel busses to achieve an overall bandwidth $O(n)$, the same as the point-to-point architecture. This gives a wire area $O(n^3\sqrt{n})$, and a power dissipation of number of wires times bandwidth times capacitance $= O(n\sqrt{n})$. Any external processor can be connected to the on-chip array via the bus, so the number of link terminals off chip is a constant at 1 (Here we ignore the fact that the bandwidth of this one terminal must increase $O(n^2)$) The Rent exponent is therefore 0, leading to a relatively low-performance architecture because in practice the bus acts as a bottleneck.

Figure 8.6 shows a network with a more regular structure in which each new processor is associated with two new links, a routing node and a network adaptor. The routing node can connect the chip to links allowing traffic to the north, east, south, and west of a processor as well as allowing traffic to pass directly between the four links. There are a total of $2n$ links, and the number of links grows $O(n)$. The bandwidth

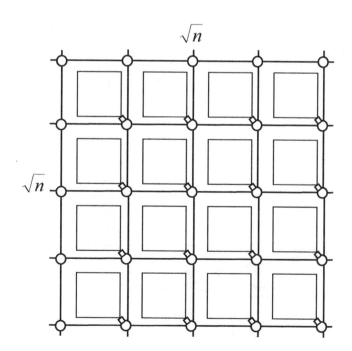

Figure 8.6 Regular network link architecture.

of each link is relatively high, because it has a unit length, independent of the array size n. The total network bandwidth is therefore $O(n)$, the total wire area is also $O(n)$ and the power is $O(n)$.

The number of chip terminals needed to extend this network off chip is $O(\sqrt{n})$, so the Rent exponent is 0.5, indicating a medium-performance architecture, where the network does not limit the performance excessively. Better performance or smaller area can be achieved by tailoring the structure to the application, for example by increasing the capacity of links likely to see large volumes of traffic, and deleting links which have no significant usage.

8.2   INTERCONNECT LINKS

The design and performance of each communication link is fundamentally affected by the fabrication technology used to implement the wire connections, and the driving gates and transistors.

Figure 8.7 shows the comparison between raw gate delay and the delay produced by a 43μ connection with each generation of fabrication technology [50]. If the minimum dimension of the process is large enough, delays are dominated by the gates, and it is possible to design a clock tree in which the clock edge arrives at nearly the same time in any part of the chip.

Below 130 nm minimum dimension, the connection delay dominates, and it is much more difficult to design a system with short enough

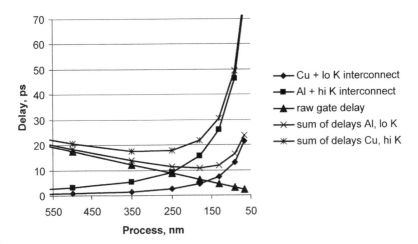

Figure 8.7   Gate delays and interconnect delays.

connections to support a single clock. At that point, increasing the processing power of a system means that more than one separately timed processor per chip is the preferred option, with some kind of regular interconnection fabric.

The characteristics of the wires making up each link also determine the performance of the network [57]. Interconnects fabricated in a submicron technology usually have a delay greater than the delay introduced by the routing nodes and they can also consume more power. The increased delay is because each connection is a CR network rather than a classical transmission line, so it is subject to dispersion which attenuates high-frequency components of the signal much more than low frequencies. Power dissipation is increased mainly because connections going long distances across a chip have high capacitance. The capacitance for wires of a constant length does not reduce with fabrication dimensions, because dielectric thickness, as well as conductor spacings are reduced at the same time as the metal width. The frequency at which this capacitance is charged and discharged is usually higher because the semiconductor devices driving the lines are operate at higher frequencies at reduced dimensions.

The metal of the interconnect line has a distributed series resistance, which is combined with the distributed parallel capacitance to ground as shown in Figure 8.8. The total resistance of the interconnect line R and the total capacitance to ground C is divided up here into many

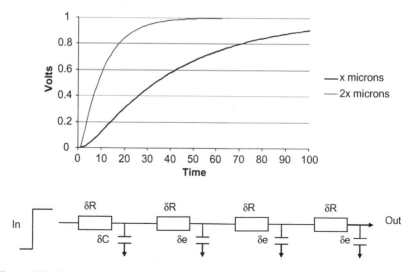

Figure 8.8 Interconnect rise time.

sections, each of which has a resistance of δR and a capacitance δC. Each section contributes to the total delay and also to an increase in the rise time of the output, but the delay is not proportional to the length of the connection. Both R and C are proportional to length x, so if the resistance per unit length of the metal used is r, then the total resistance is $R = xr$, and the total capacitance is $C = xc$ where c is the capacitance per unit length. The result is that the delay is proportional to x^2. Typical rise times are shown in Figure 8.8 for interconnects of length x and $2x$ where x is approximately 5, and the time scale is in units of rc. The delay can be seen to be about 8 units for the shorter wire and 32 units for the longer one. Similarly the output waveform is always a factor of four slower at any point than the longer wire. The result is that both the latency and the bandwidth of data that can be transmitted along any interconnect worsens according to the interconnect length squared.

The power required to transmit data is also increased because of the relatively high capacitance of long interconnects. Each transition from a low voltage level to a high requires the capacitance to be charged from the power supply, usually through the resistance of a p-type transistor, so the energy taken from the power supply is $QV = CV^2$, where Q is the charge required to change the interconnect voltage. Half of this energy is dissipated in the p-type transistor, and half is stored in C. The next transition, from high to low, involves discharging the capacitance, usually through an n-type transistor to ground and thus dissipating the remaining energy. For the pair of transitions a total energy of CV^2 is irrecoverably lost, so it is important to ensure that the maximum amount of data is carried for each pair of transitions.

A simple way of improving the bandwidth of an interconnect line is shown in Figure 8.9, where the line is divided up into n sections separated by inverters which restore the rise time at the output of the inverter.

Now if we assume that the inverter output rise or fall time is much faster than its input rise or fall time, the rise time of one section of the line, is determined by the line alone as in Figure 8.8, the total line delay is improved to $RC/n + nt_d$, where t_d is the delay of an inverter and RC/n^2

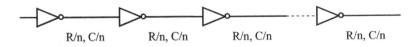

Figure 8.9   Inserting inverters.

is the delay of one section. The optimum number of inverters depends on the ratio of inverter delay to interconnect delay and is

$$n = \sqrt{\frac{RC}{t_d}}$$

For processes with large gate delays and small rc fewer inverters will be needed than for processes with small gate delays and large rc. Rise time, and therefore bandwidth is also improved by the re-standardization of the rise time at each inverter stage. The power dissipation for the combination of inverters and interconnect sections is greater than for the undivided interconnect, but only because of additional capacitance associated with each inverter. If the added capacitance due to the inverters is much smaller than C, the power dissipation would not be significantly changed if same transition frequency is kept, but, of course, the objective of including the inverters is to increase the frequency.

The bandwidth of a link can be further improved by pipelining. Flip-flops can be used instead of inverters, so that a pipeline is created as in Figure 8.10. In the pipeline each flip-flop can have a different state so there can be n transitions in flight between the ends of the link.

The delay between input and output is now n cycles of the clock, rather than the transmission delay of the rising or falling edges, but the number of possible transitions is one per cycle, which can lead to better bandwidth because the clock cycle time can be less than the total delay. Power dissipation in this pipelined scheme is significantly increased because the number of transitions per second is increased. Since the number of transitions per second is directly related to the bandwidth, both the power dissipation and bandwidth are increased at the same rate.

Long interconnects in submicron processes are also subject to crosstalk caused by the placement of one interconnect near another.

If two metal interconnect structures carrying different signals are long, and close to each other, the capacitance between them can be significant. A rising edge on one line can cause a positive pulse on the other because of the capacitive coupling, as in Figure 8.11. In the worst case this may cause a gate on the victim wire to change state, and this can even cause a system failure if a latch gets set or reset as a result. In

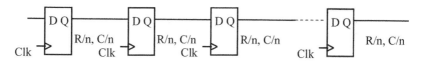

Figure 8.10 Pipelined link.

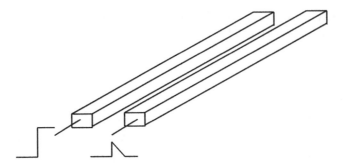

Figure 8.11 Cross-talk.

practice the effect is usually for the aggressor signal either to assist a rising edge on the other line, or to oppose it if it is a falling edge. In either case the effective delay on the victim line is altered.

Transmission of data along a long interconnect path can be done by serial or parallel methods, and the relative merits of each methods can depend on the technology of the implementation and the needs of the application. An issue for high-speed communication is the effect of a transition on one line on the following transition on the same line. All positive going transitions must be followed by a negative going one, and vice versa, so if a positive transition has not quite reached the high voltage level at the output before the following negative going transition starts, the negative transition will start from a lower point.

This effect is illustrated in Figure 8.12, which shows that a long period at a low voltage level followed by a series of successive positive

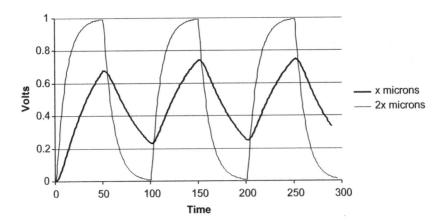

Figure 8.12 Baseline shift.

and negative transitions can cause a gradual shift in the starting point of each transition. This is sometimes known as 'baseline shift' and can limit the bandwidth of data transmission. The first pair of a burst of closely spaced transitions can be missed after a long quiet period. Short interconnections are much less affected than long interconnections because their rise time is much faster, and low bit transmission rates are much less affected than high bit rates because there is sufficient time for the line to recover between transitions.

8.3   SERIAL LINKS

8.3.1   Using One Line

The simplest way of transmitting data in a network on silicon is to use a serial link, and the simplest form of serial link uses a single line with the data coded as high for a 1 and low for a 0. Figure 8.13 shows the resulting NRZ code in which transitions only occur if the data in successive bits changes. On average there will then be one transition every two bits transmitted, because there is an equal probability of a change or no change between bits. In order to recover the data it is necessary to know the time between bits, known as the bit period; unless this time is somehow communicated between the sender and the receiver, the data cannot be recovered.

The initials NRZ stand for non-return to zero, because the voltage level is always either high, or low and never returns to a rest level halfway in between. Codes that do this are called return to zero or RZ. A scheme with similar characteristics to NRZ is NRZI, Figure 8.14

In NRZI (non-return to zero invert) a transition is inserted for a 1 on the boundary of the bit period, but not for a zero. Again the average number of transitions per bit is 1/2, and the voltage has only two levels, high and low. NRZI is the more frequently used code for example for

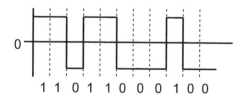

Figure 8.13   NRZ coding.

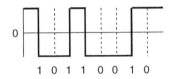

1 0 1 1 0 0 1 0

Figure 8.14 NRZI code.

the USB connection on a PC and other common serial paths. The major problem with both NRZ and NRZI in an asynchronous network is clock recovery. There is nothing in the code which defines the time of the bit period and the correct recovery of the information requires an accurately timed clock to define when the bits are sampled. Writer and reader are usually some distance away and their clocks may be unrelated. Both NRZ and NRZI can have long periods with no transitions which means that it is difficult to recover the clock from the data. In NRZ the sequence 00000 or 11111 would result in no transitions as would 00000 in NRZI. Any clock recovery circuit in the reader has to rely on at least one transition appearing at certain minimum rate to be able to accurately predict all the bit times between those transitions. Usually this means a phase-locked loop (PLL) which does not change its frequency over many cycles is needed for clock recovery.

A code which does provide frequent extra transitions is Manchester Code (Figure 8.15) which is a simple form of phase encoding.

In this code there is always a transition in the centre of the bit cell, a rising transition for a 0 and falling transition for a 1. An additional transition has to be inserted on the boundary of the bit cell if two successive bits have the same value, rising for 1,1 and falling for 0,0. The average number of transitions per bit is 1.5, but there can be as many as 2. Despite the fact that the time between transitions can be half the value of the previous two codes, the baseline shift is not large because

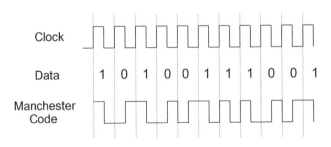

Clock

Data   1 0 1 0 0 1 1 1 0 0 1

Manchester
Code

Figure 8.15 Manchester Code.

the maximum time at a high or low voltage level is one bit period, and the minimum time half a bit period so there can never be a long sequence with no transitions. Clock recovery is relatively easy.

8.3.2   Using Two Lines

Incorporating the clock into the coding on a single line can be done either with a large number of clock transitions per bit, for example Manchester Code, where the number of transitions per bit is high so that the data bandwidth (number of data bits transmitted per second) suffers, or by only having a few transitions dedicated to the clock, in which case clock recovery can be difficult. Using two lines in a serial link allows both clock and data to be transmitted without the need for any clock recovery system. A simple two-line solution is the dual-rail scheme (Figure 8.16) in which a rising transition on line 1 represents a 1 and a rising transition on line 0 represents a 0.

Only one line can have a high level at any time and there must be one transition on one of the lines each bit period, so the transition itself provides the timing for the arrival of data. The line is returned low when the data has been accepted, and both lines low represents a 'spacer' between the data bits. Dual rail is well suited to self-timed systems and delay insensitive systems in which there is a separate path with a high-going transition to acknowledge the arrival of data before the spacer can be transmitted. In a delay insensitive system the spacer is acknowledged by returning the acknowledge signal to low before a new data item can be sent. Dual-rail protocols use at least two transitions per bit of data transmitted, and also need an acknowledgement transition to be fully self-timed.

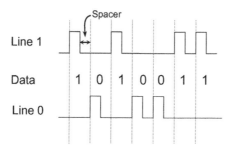

Figure 8.16   Dual-rail level signalling.

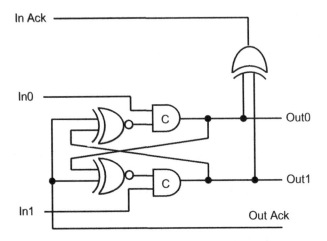

Figure 8.17 Dual-rail transition signalling.

It is possible to reduce the number of transitions in a dual rail protocol by using transition signaling as in Figure 8.17. A transition (high-going or low-going) on one line represents a 1 and a transition on the other line represents a 0. Initially both inputs and both outputs in Figure 8.17 are low, and the other inputs to the C gates are high. If In0 goes high, representing the arrival of a 0, the corresponding C gate will switch, and Out0 goes high, so an input acknowledgement transition is sent on In Ack. When the output data has been received and stored in the receiver, it is acknowledged by a transition on Out Ack. The two inputs to both C gates are now different again, that is, In0 is now high, and the other C gate input is low. Similarly In1 is still low and the other input to the second C gate is still high. Any further input transition on either input causes the corresponding output to produce a transition and an input acknowledgement. The penalty for transition reduction is a rather more complex receiver/transmitter circuit.

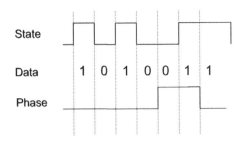

Figure 8.18 Level encoded dual-rail.

Table 8.1 Summary of coding schemes for a serial link.

	Transitions per bit	Bandwidth	Lines	Clock recovery/ timing	Baseline shift
NRZ	1	High	1	May need a PLL	Yes
NRZI	1	High	1	May need a PLL	Yes
Manchester	1–2	Low	1	Simple clocking	Low
Dual-rail	2	Low	2	Self-timed	Yes
Dual-rail Transition	1	High	2	Self-timed	Yes
LEDR	1	High	2	Self-timed	Yes

A variant of dual rail is level encoded dual-rail (LEDR), Figure 8.18, [59] in which the two lines are encoded as state and phase bits. The bit sequence $B(i)$ is mapped directly into the state bit $S(i)$, that is, $S(i) = B(i)$ for all i. To ensure one transition for every bit, the phase bit changes if $B(i) = B(i-1)$, but does not change if $B(i) <> B(i-1)$.

Since there is only one transition per bit and no acknowledge signal LEDR is potentially lower power than other forms of dual-rail, and is also capable of high data rates since there is always a full bit period between transitions. It may, however suffer from baseline shift at high data rates because there can be long periods without a transition on either the state or the phase lines.

Encoding and decoding LEDR is simple since to encode the relationships $S(i) = B(i)$, and $P(i) = NOT(B(i)$ XOR $P(i$-1) XOR $B(i$-1)) can be used. For decoding the clock is $S(i)$ XOR $P(i)$, and $B(i) = S(i)$

There are other possible encoding schemes for a single bitstream serial link, but the ones given above are representative of the options that can be considered, and they are compared in Table 8.1. Protocols with the lowest number of transitions per bit and highest potential bandwidth are NRZ, NRZI, and LEDR. Of those LEDR uses more lines, and therefore may occupy more silicon area, but clock recovery is easier.

8.4 DIFFERENTIAL SIGNALLING

At the physical level, each signal can be carried by a single interconnect line, or by a pair of lines using the difference in voltage or current between the lines to represent a 1 or a 0, [58]. The circuit of a differential link is shown in Figure 8.19.

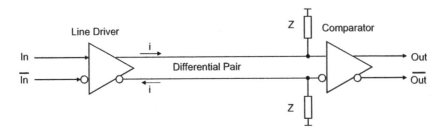

Figure 8.19 Differential link.

In this case a line driver is used to drive a defined current i in one direction in one of the pair of interconnecting lines, and the opposite direction in the other. Each line is terminated by an impedance Z at the far end equal to the characteristic impedance of the transmission line formed. The voltage developed across the pair is used as the input to a comparator which outputs levels corresponding to the original inputs.

One advantage of this arrangement over a single-ended transmission line is that it is less susceptible to external electromagnetic interference, because both lines are likely to be affected similarly. Interference, either from power supply noise, or external fields normally gives common mode noise, but the signal itself is differential. The receiving comparator is designed to reject common mode noise effects, and is sensitive to differential mode inputs. A balanced line also produces less external field itself, thus reducing the effect of any noise from the signal being transmitted to other connections. The terminations have a typical impedance of around 100 Ω, designed to match the characteristic impedance of a typical line pair, and the value of I is 0.1–1 mA. The low value of Z means that the RC time constant of the connection is much lower than without the termination. C remains the same but R effectively appears in parallel with Z. Power on the other hand is considerably higher because both line driver and comparator may have standing current flowing in them, and a typical connection can dissipate tens of milliwatts. Transitions in the link are a much lower proportion of the power dissipated because the voltage swings are only of the order of 10 – 100 mV. The small size of the voltage swing means that the input transistors may need to be large to minimize any input transistor parameter variability.

A differential link can be effective in a high bandwidth application where power dissipation is less important. Achieving the maximum bandwidth involves reducing the cross-talk, and also shielding the link from other aggressor lines. An LEDR scheme in which the lines are

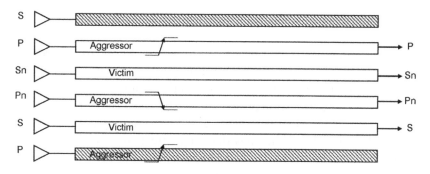

Figure 8.20   Minimizing cross-talk in LEDR.

interleaved to minimize crosstalk, and also provide shielding is shown in Figure 8.20, [58]. In this example the P and S lines drive two lines each and the inverse of P and S, Pn and Sn, drive one each. The lines are positioned so that each of the four inner lines is surrounded by a signal and its inverse, so that the two aggressor lines on either side of a potential victim carry the true and inverse of the same signal. Pn is surrounded by S and Sn, Sn by P and Pn etc. The cross-talk from the aggressors is cancelled out by this arrangement. The two outside lines are not used as outputs so act only as shields for the four inner lines.

Because of the noise cancellation properties of the layout, lines can be placed quite close together when compared with lines carrying unrelated signals.

8.5   PARALLEL LINKS

Parallel links can have a higher bandwidth than serial links simply because there are more lines to carry information at one time. NRZ or NRZI representations could be used on each line in an n line parallel link where 2^n symbols can be transmitted in each bit period. Unfortunately in a parallel system the clock recovery problems of these codes are made worse by the additional skew on each line. Bit timing on each line varies as a result of line length, process variation causing metal thickness and capacitance to differ on each line, and cross-talk between lines which can assist or slow each edge. Self-timed methods are far easier to implement, especially in a globally asynchronous environment with multiple processor clocks, so codes which allow for the self timing of each symbol are more useful.

8.5.1   One Hot Codes

Dual-rail coding is an example of a self-timed communications protocol in which the presence of a symbol, in this case 1 or 0, is indicated by one of the two lines having a high level. When the symbol is acknowledged, both lines become low, indicating a spacer between the current symbol and the next. The next time one line goes high a new symbol is present. Skew in the delay between lines does not affect the time of arrival of each signal, because only one line changes state for each new symbol, and it changes back when the symbol is acknowledged. Only one of the two lines can be high at the same time, so dual rail is known as a 'one hot' code. One hot codes do not need to be restricted to two lines. Three, four, five or any number of lines can be used, and provided only one line changes state when a symbol is transmitted or acknowledged the exact timing is given by one transition and cannot be affected by skew. The number of symbols that can be transmitted when one line out of n goes high is n in an n hot code.

An example of a one of four code is shown in Figure 8.21. Here when one line goes high, a new symbol is present, and when the symbol is acknowledged the line returns low, so that the state with all lines low represents a spacer. Since there are four lines, the number of possible symbols that can be transmitted in one bit period is four, which might be the symbols 0,1,2, and 3.

An implementation of transmit/receive circuits in a one of four coded link known as CHAIN [60] is shown in Figure 8.22. The corresponding waveforms for this circuit are given in Figure 8.23.

The inputs in Figure 8.22 start in the spacer state with In0 ... In3 all low, and In Ack high. One of the four input signals In0 ... In3 goes high

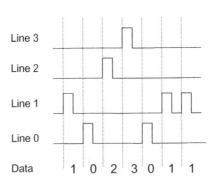

Figure 8.21   One of four code.

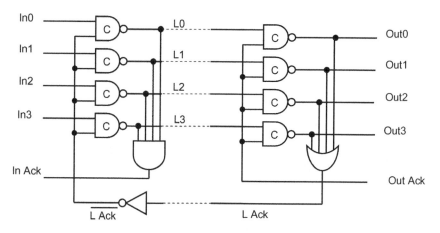

Figure 8.22 One of four CHAIN transmitter and receiver links.

to indicate that a symbol (0,1,2,or 3) has arrived at the input. The signal L Ack is low, and its inverse high, so one of the C gate outputs will now go low driving one of L0 ... L3 low and making In Ack low a little later. In Ack acknowledges that the input has arrived on L0 ... L3. This input acknowledgement causes the high input to be lowered later, as indicated by the dashed arrow. When Out Ack goes low indicating that the outputs can change, one of the four output signals Out0 ... Out3 goes high, L Ack becomes high and its inverse goes low.

When the high input returns low to produce the spacer, L0 ... L3 become all high, and InAck returns high. Out0 ... Out3 can now return to the spacer state when Out Ack goes high.

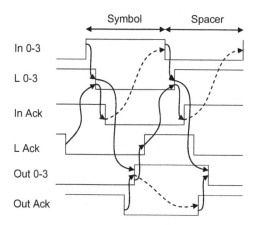

Figure 8.23 One of four waveforms.

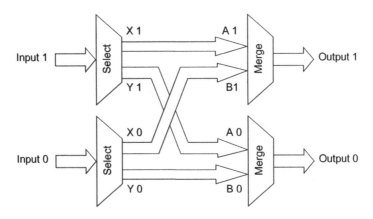

Figure 8.24   Routing with select and merge.

In these circuits, the C gates act as latches, holding the input symbol on L until L Ack goes high, and holding the output symbol on Out until Out Ack goes high. As a result the link can be split into a number of asynchronous pipelined sections, reducing the effect of cross-talk and poor rise times due to long line length.

In a network on chip the links transmit data between processors and routing nodes. The routing nodes determine which path each packet must follow, and each routing node has the capability to direct a packet from an import port to the correct output port.

A very simple routing node is shown in Figure 8.24 which allows either of the two input ports to be directed to either of the two output ports. On the left-hand side of the figure the select blocks have two output paths, one to each merge block. One of these is selected, and the merge blocks pass the active input to the correct output ports. More complex routing nodes like can be built using select and merge circuits, and in particular the node shown in Figure 8.3 can be built from two of the simpler nodes shown in Figure 8.24.

Figure 8.25 shows the circuit of a select block designed for the one of four code. The two select signals, Sel X and Sel Y enable the upper and lower paths respectively. A valid input causes one of the four outputs on the selected path to go high, with all the outputs on the other path remaining low. The input is acknowledged when In Ack goes high.

The circuit of the merge block is shown in Figure 8.26. If the previous symbol has been acknowledged, a new valid symbol on either the A or the B inputs to the merge block causes a new output, which is acknowledged itself by AB Ack.

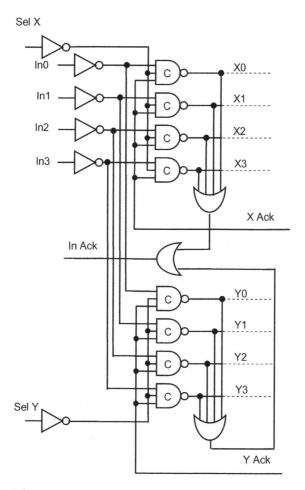

Figure 8.25 Select.

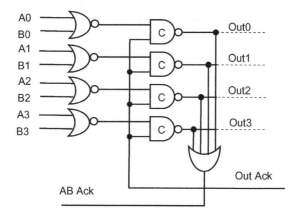

Figure 8.26 Merge.

8.5.2   Transition Signalling

A one hot code based on level signaling needs two transitions to transmit a symbol, plus any acknowledge transitions that may be required for control of the pipeline, but the power required remains constant for one of n, no matter what the value of n because only one line changes at any time. The area overhead in terms of number of lines however, increases in with n, as does the number of symbols transmitted for each request–acknowledge cycle. Some of the trade-offs are illustrated in Table 8.2.

In order to make a comparison, we can compare a one of four level code, which needs five lines, with five lines used in a link in which four transmit NRZ data, and one is used for the clock timing. The total number of symbols that could be transmitted by the four NRZ lines is 2^4, or 16. On average there would be a probability of 0.5 of a transition on each line between symbols, giving 2 transitions for four lines plus one transition for the clock. A one of four link also needs five lines and uses 2 transitions per symbol plus two transitions for control. It would require more power, but it is completely self-timed, is unaffected by skew, and does not need clock recovery circuits. The design is much easier, and the bandwidth not much less than the NRZ case because skew and delay can be ignored. The one hot codes shown in Table 8.2 use level signalling, bit it is possible to use a one of four transition signaling approach in which each symbol is represented by a transition on one of the lines.

Figure 8.27 shows how this is done. The circuit is initialized so that both inputs to each C gate are different. When an input transition arrives, it causes the corresponding C gate output to change. A four input XOR gate then gives an input acknowledgement transition, and when the output is acknowledged all C gate inputs again become different. While this scheme uses a single transition to carry each event, and there are fewer transitions per symbol, there are a large number of relatively slow, complex XOR gates. A smaller number of transitions per bit period should lead to a greater bandwidth, but because of the

Table 8.2   One hot codes.

	Transitions per symbol	Timing control transitions	Symbols	Bits	Lines
NRZ, four lines	2	1	16	4	5
Dual-rail	1	1	2	1	3
Transition					
1 of 2 level	2	2	2	1	3
1 of 4 level	2	2	4	2	5
1 of 8 level	2	2	8	3	9

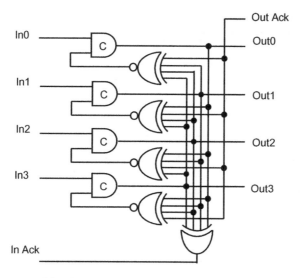

Figure 8.27 One of four transition signalling.

complex circuitry, the bandwidth of the link is not necessarily any better than the one of four scheme using level signaling.

8.5.3 *n* of *m* Codes

An obvious extension of the idea of the one out of *n* self-timed link where a symbol is represented by a single 1, and a spacer by all 0, is to use *n* out of *m*. Since each line can have a 1 or a 0 on it, if there are *m* lines the number of combinations in which *n* lines are 1 and *m − n* lines are 0, is

$$\frac{m!}{n!(m-n)!}$$

Each of these can represent a different symbol, giving four symbols for $n = 1$ and $m = 4$, and 20 symbols for a three out of six code, where $n = 3$ and $m = 6$, and in general many more symbols for a given number of lines. We can make a self-timed link by choosing the code all 0 to represent the spacer, as before, so that in an 3 out of 6 code, 3 transitions indicate the arrival of a symbol, and the reversal of those 3 gets back to the spacer. A similar circuit to Figure 8.22 could be used for a pipelined 3 out of 6 repeater, but with the AND and OR gates replaced by circuits to detect three 1s or six 0s.

Table 8.3 shows some possible codes in which the number of bits transmitted per line varies from 0.4 for the 1 of 4 code to 0.57 (3 of 6), and the number of transitions per bit from 1.2 (2 of 9) to 2 (1 of 4). Most of them

Table 8.3   N of *m* codes.

	Transitions per symbol	Number of symbols	Bits	Lines
1 of 4	2 + 2	4	2	5
3 of 6	6 + 2	20	4	7
2 of 7	4 + 2	35	4	8
3 of 8	6 + 2	56	5	9
2 of 9	4 + 2	36	5	10

can transmit a number of symbols which is not an exact power of two, so, for example, a 2 of 7 code could transmit $2^5 = 32$ data symbols and another 3 symbols which could be used to indicate the start or end of a packet.

8.5.4   Phase Encoding

The number of symbols transmitted on a link with a given number of lines can be increased further by using the order in which the transitions arrive [60] rather than just their presence or absence. If there are two lines with a transition on each line for every symbol, there are just two possibilities for changing the order. The transition on line a is first, or the transition on line b is first. Two symbols can be transmitted, 0 and 1, and the protocol is self timed, because the symbol is not complete until both transitions have occurred.

In Figure 8.28 the waveforms show the relative timing between line a and line b, which indicate whether a 0 or a 1 is being transmitted. In this code there are always two transitions per bit, compared with the one transition per bit for the data, and one for the request/acknowledge mechanism in dual-rail transition coding. Because the data is carried by timing variations rather than different levels, it is a form of phase encoding.

One advantage of this form of coding is its relative insensitivity to transient errors. Transient errors due to cross-talk, or environmental

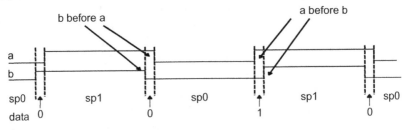

Figure 8.28   Dual-rail phase encoding. Reproduced from Figure 1, "Multiple-Rail Phase-Encoding for NoC", by C. D'Alessandro, D, Shang, A. Bystrov, A. Yakovlev, and O. Maevsky which appeared in Proc. ASYNC'06, Grenoble, France, March 2006, pp. 107–116 © 2006 IEEE.

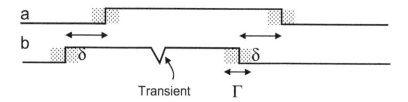

Figure 8.29 Robustness to transients. Reproduced from Figure 1, "Multiple-Rail Phase-Encoding for NoC", by C. D'Alessandro, D, Shang, A. Bystrov, A. Yakovlev, and O. Maevsky which appeared in Proc. ASYNC'06, Grenoble, France, March 2006, pp. 107–116 © 2006 IEEE.

interference increase as the dimensions of the fabrication process reduce, and level-based coding methods can be vulnerable to hazards such as a transient change in the level of one line. The fact that valid data is recorded only when both lines have changed state in dual rail phase encoding means that a hazard on one line will not be recorded as data.

The time between transitions, δ, must always be greater than any jitter or noise Γ in the positioning of the transitions introduced by the link for data to be correctly recovered, but a transient in the spacer region, as shown in Figure 8.29 is filtered out because the next symbol is not complete until both lines have changed state. Provided the time of the spacer is greater than the time needed to assemble the symbol δ, the protocol is relatively immune from transients.

Phase encoding can be extended to as many lines as is required, and Figure 8.30 shows a four-line system with transitions a, b, c and d, ordered in time. Each of the permutations, abcd, abdc, adbc, etc. represents a different symbol. In Figure 8.30 the reference signal simply shows the start time for the encoding of any symbol, it does not need to be transmitted unless some redundancy is required for fault tolerance. Without the reference signal deadlock is possible if one of the transitions is missing, but this could be detected if the reference for the next signal is received before the symbol is complete.

The number of possible symbols that can be transmitted at one time increases rapidly as the number of lines increases. The number of possible permutations of transitions is $n!$. where n is the number of lines. Two lines give two symbols, three give six symbols and four give 24 symbols, so if the transmission of a p bit packet is considered, the number of transitions per packet is

$$\frac{p}{\log_2(n!)} n$$

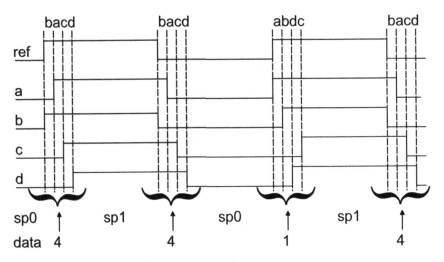

Figure 8.30   Multiple-rail phase encoding.

where n is the number of lines. For an n of m level encoding scheme we would have

$$\frac{p}{\log_2\left[\dfrac{m!}{n!(m-n)!}\right]} 2n$$

transitions per bit.

Phase encoding is compared with n of m codes for a 128 bit packet in Table 8.4. For the same number of lines, a phase encoding protocol allows many more symbols, than an n of m protocol. The number of transitions used for n of m depends on whether level or transition coding is used, here the comparison is made with level coding because this results in simpler encoding and decoding circuits for n of m. From the point of view of the code characteristics, with a large number of lines, a packet can be transmitted quicker using phase encoding than for n of m encoding because of the larger number of bits per symbol.

Table 8.4   Comparison of n of m and phase encoding for a 128 bit packet.

	Possible symbols	Bits/ symbol	Extra states	Transitions per 128 bit packet	Symbols/ packet
4-rail phase encoding	24	4	8	128	32
6-rail phase encoding	720	9	208	90	15
1 of 4 level	4	2	0	128	64
3 of 6 level	20	4	4	192	32

The bandwidth of a link depends not only on the code characteristics, but also the quality of the link itself. Poor-quality links in which the one edge can overtake another because of cross-talk effects will have a poorer bandwidth than high quality links. The issues that must be considered include:

- the minimum resolvable spacing between transitions δ;
- the time necessary for the receive circuits to decode each symbol (which may depend on δ);
- the noise, jitter, and crosstalk effects introduced by the link, Γ.

If we combine these factors into a link quality factor χ, which is the minimum proportion of δ reaching the receiver, then as χ deteriorates, the time between symbols must increase to compensate by increasing δ and allowing the receiver circuits more time. If the value of χ is much below 0.5, the bit rate will be significantly affected. For a typical link where $\chi = 0.8$, as we increase the number of lines, the receiver complexity increases significantly.

This reduces the rate at which symbols can be decoded faster than the number of bits per symbol increases. This effect can be seen in Figure 8.31 where the frequency of symbols falls, and the bit rate reaches a peak at around 8 lines.

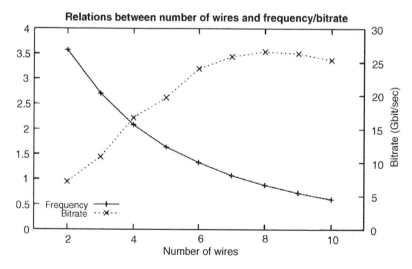

Figure 8.31 Link bit rate against number of lines. Reproduced from Figure 6, "Multiple-Rail Phase-Encoding for NoC", by C. D'Alessandro, D, Shang, A. Bystrov, A. Yakovlev, and O. Maevsky which appeared in Proc. ASYNC'06, Grenoble, France, March 2006, pp. 107–116 © 2006 IEEE.

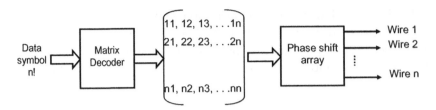

Figure 8.32 Block diagram of sender. Reproduced from Figure 7, "Multiple-Rail Phase-Encoding for NoC", by C. D'Alessandro, D, Shang, A. Bystrov, A. Yakovlev, and O. Maevsky which appeared in Proc. ASYNC'06, Grenoble, France, March 2006, pp. 107–116 © 2006 IEEE.

8.5.4.1   Phase encoding sender

The sender for a phase encoding scheme needs to encode the data into a sequence of transitions, the order of which represents the symbol. Figure 8.32 shows a block diagram of a sender for n lines.

The sender encodes a symbol in an $n \times n$ matrix M whose elements correspond to internal control signals. These signals are used to control the phase encoding logic and are such that each row of the matrix corresponds to a line and each column to a possible delay achievable on the lines. The n-line phase encoding logic encodes each of the $n!$ states into a symbol-dependent matrix which determines the delays.

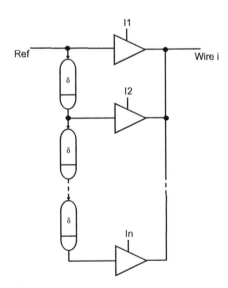

Figure 8.33 Phase shift array. Reproduced from Figure 7, "Multiple-Rail Phase-Encoding for NoC", by C. D'Alessandro, D, Shang, A. Bystrov, A. Yakovlev, and O. Maevsky which appeared in Proc. ASYNC'06, Grenoble, France, March 2006, pp. 107–116 © 2006 IEEE.

As an example, 4-line phase encoding logic can encode 24 different symbols which we indicated by the numerals [1…24]; If we wish to send the symbol corresponding to the numeral 17, the corresponding matrix

$$M(17) = \begin{bmatrix} 0 & 0 & 0 & 1 \\ 0 & 0 & 1 & 0 \\ 1 & 0 & 0 & 0 \\ 0 & 1 & 0 & 0 \end{bmatrix}$$

is generated. Since the lines with letters a–d corresponding to, respectively, rows 1–4 the sequence will be cdba.

The phase encoding logic, Figure 8.33, consists of a delay chain whose input is the reference signal and outputs are the n differentially delayed versions of the reference; it also has an array of tri-state buffers to direct the appropriate delayed version of the reference to the corresponding line.

8.5.4.2 Receiver

The receiver for such a system is more complex than the sender, as it needs to identify and decode the sequence of transitions on n lines. Mutual exclusion elements (MUTEXs) are employed to arbitrate between two requests and the minimum time between transitions is arranged so that pairs of transitions very close in time to each other do not drive the device into metastability. The receiver design shown in Figure 8.34 uses a MUTEX for each pair-wise combination of lines; this array of MUTEXs produces a binary output which is unique for each sequence and which is then fed to a decoder to extract the data, and the decoding array is shown in Figure 8.35.

In this approach, an array of $n(n - 1)/2$ MUTEXs is required for n lines, to cover all possible pair-wise combinations if only rising or

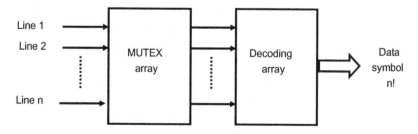

Figure 8.34 Receiver block diagram. Reproduced from Figure 8, "Multiple-Rail Phase-Encoding for NoC", by C. D'Alessandro, D, Shang, A. Bystrov, A. Yakovlev, and O. Maevsky which appeared in Proc. ASYNC'06, Grenoble, France, March 2006, pp. 107–116 © 2006 IEEE.

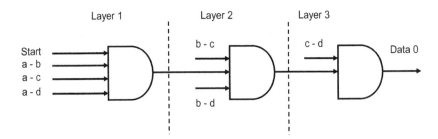

Figure 8.35 Decoding array logic layers. Reproduced from Figure 8, "Multiple-Rail Phase-Encoding for NoC", by C. D'Alessandro, D, Shang, A. Bystrov, A. Yakovlev, and O. Maevsky which appeared in Proc. ASYNC'06, Grenoble, France, March 2006, pp. 107–116 © 2006 IEEE.

falling transitions are used; or in the case of both transitions being used the number of MUTEXs doubles.

The logic used to decode the output of the MUTEX array is organized into layers, as shown in Figure 8.35, each identifying the line which wins a number of arbitrations. The sequence can be reconstructed by analysing the number of arbitrations each line wins. The first line will win all arbitrations with all other lines; the number of these arbitrations is thus $n - 1$ for the first line. The second line will win all arbitrations with the other lines apart from the first and so on. Therefore, each layer identifies the line which wins $n - l$ arbitrations: l is both the layer and the position of the line in the sequence. This means that the number of logic layers is $n - 1$, as the last line will not win any arbitrations and can be identified by eliminating all the previous lines.

This approach has the advantage of reducing the size of the logic compared with the other possibility of a tree of arbiters; it also has the additional advantage of reducing the latency of the receiver, as only one layer of arbiters is used:

The MUTEX array is implemented including memory logic around the MUTEXs to achieve fault tolerance and guarantee correct operations. Figure 8.36 shows the implementation of each element in the array, in particular the element responsible for arbitration between lines a and b. Apart from the MUTEX in each element, there are two AND–OR gates, two three-input C-gates and one two-input OR gate. The two AND–OR gates are used together with the C-gates to construct memory elements to 'keep' the outputs of the MUTEX. When one line wins arbitration and all other lines have switched, the corresponding AND–OR gate output keeps the output of the MUTEX stable until the C-gate changes state. This will only happen when all lines change state (at the next symbol). This mechanism guarantees that the output of the array is stable between symbols and will not produce an error for transient faults occurring in the spacer.

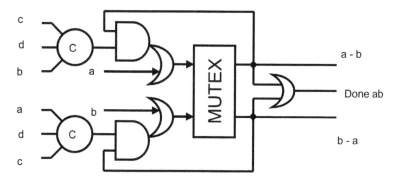

Figure 8.36 Fault-tolerant MUTEX.

Because there is a memory in the circuit of Figure 8.36, completion detection of arbitration is quite simple: only one OR gate for each element is enough because, after resolving, only one output of the MUTEX will be logic high, whilst in idle mode the outputs will be both logic low.

8.5.5 Time Encoding

The number of symbols that can be encoded on a set of n lines can be increased further by encoding the time of the transitions rather than the sequence of transitions. Consider the case of two lines in Figure 8.37.

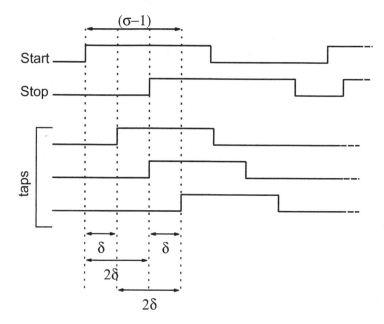

Figure 8.37 Two-line time encoding.

One line transmits a start transition and the other a stop transition, the time between the two transitions representing the symbol. In Figure 8.37 the time is measured in multiples of a minimum spacing δ whose value is determined by considerations similar to those of the previous section. If the spacing between symbols is a time $\sigma\delta$, where σ is an integer, then there will be $\sigma-1$ possible times when either the start transition or the stop transition can be placed. Some of these combinations lead to the same time between the two transitions (2δ in the figure). If we always place the start before the stop, there are $\sigma-1$ possibilities, not including 0, and if we use negative times, where the stop can be before the start or at the same time as the start, there are another σ possibilities. The number of bits that can be transmitted in the time $\sigma\delta$ is there fore either $\log_2 \sigma - 1$ or $\log_2 (2\sigma - 1)$. Encoding is relatively simple, a reference transition, which could be the same as the start transition if only positive times are allowed, is delayed by a tapped delay line with taps at increments of the time δ, and the tap giving the time corresponding to the correct symbol is selected for the stop transition. In the two-line protocol shown, $\sigma = 5$ gives four possible time slots and allows two bits to be transmitted per symbol, twice the rate possible for other dual-rail protocols. The number of transitions per bit for this protocol would be one, as good as dual-rail transition signalling on two lines.

Decoding is done by means of a time to digital converter (TDC). Many designs for TDCs exist [62–64] and two of the simplest are shown in Figure 8.38.

Figure 8.38(a) shows a TDC made from a tapped delay line and a column of MUTEXs. The time of the start signal is compared with the stop signal by the first MUTEX. If the start is before the stop Out 0 will be set low. For the second MUTEX the start signal delayed by a time δ is compared with the stop signal, so Out 1 is set low if the stop transition is less than δ later than the start. In general if the time between start and stop is between $m\delta$ and $(m+1)\delta$, the MUTEX outputs Out 0 to Out $m - 1$ will be set low, and outputs Out m to Out n will be high. The position of the change in output value between outputs, is therefore a unique identifier for the decoded symbol.

Often the delay elements are made from a pair of inverters, and so the smallest value of δ that can be used is about the same as a gate delay, and this can limit the time step that is resolvable by the TDC. Consequently the bandwidth of the link is limited. An alternative which allows a much smaller value of δ is shown in Figure 8.38(b). Here the delay difference in the start and stop signal paths is determined by a slightly smaller delay in the stop signal path for each MUTEX than in

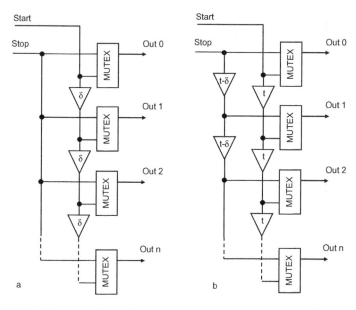

Figure 8.38 (a) Time to digital converter; (b) Vernier TDC.

the corresponding start signal path. This delay difference can easily be made much smaller than the delay in a pair of inverters, and so values of δ an order of magnitude less than an inverter pair can be used.

As the number of possible time slots, σ, in a two-wire time encoded protocol increases, the number of symbols that can be transmitted increases in proportion, but the number of bits per symbol only increases as $\log_2 \sigma - 1$. The symbol frequency reduces with σ because not only does the maximum time between start and stop transitions increase, but the complexity of the encode–decode circuits also increases, causing more delay. In Figure 8.39 the symbol frequency and bit rate are plotted against σ. An optimum in terms of bit rate is reached at $\sigma = 5$. Beyond that the gain in bits per symbol is more than offset by the drop in signal frequency. Since $\sigma - 1$ is not a power of two for $\sigma = 6, 7$, and 8, in fact only 2 bits per symbol are be encoded in an eight-slot protocol, and there are peaks in the bit rate shown by Figure 8.39 at $\sigma = 5$ and $\sigma = 9$. Between the peaks there are extra symbols available in the protocol which are effectively lost to the bit rate measure being used.

In a parallel link based on time encoding the number of possible symbols is greatly increased. If the first line carries a start reference, and all the others are stop lines, then using only positive time differences we will get $\log_2 \sigma - 1$ bits for every stop line.

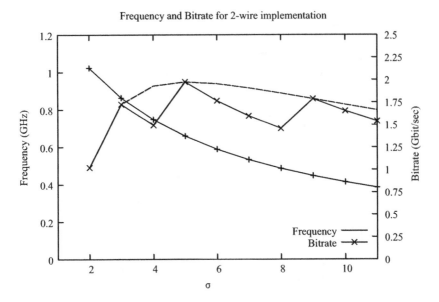

Figure 8.39   Bit rate for two-line time encoding.

In a five-line parallel time encoded protocol it should be possible to get up to four times the bit rate of a two-line protocol, but this is not possible in practice because of a number of factors. Amongst these, mismatch in process variations can cause a systematic delay offset to appear between any two lines, which could cause errors in decoding, and cross-talk causes both transient delay/phase error and symbol-dependent delay corruption

As the wires are always 'allies' in terms of cross-talk (for small time separations), the longer the wire, the more corrupted the phase relationship between the wires. The result in terms of bit rate for a five-line protocol is shown in Figure 8.40, where again the peak rate is between $\sigma = 5$ and $\sigma = 9$, with a bit rate of between two and three times the two-line protocol.

One of the properties of time encoding is that the symbol time does not need to be fixed. A two-line time encoded protocol with $\sigma = 5$ has a start transition at time 0, and can transmit symbols with the stop transition at times δ, 2δ, 3δ or 4δ. Only the symbol with a stop transition at 4δ needs to have the next start transition at time 5δ, the others could be followed by a new symbol at 2δ, 3δ, and 4δ.

Figure 8.41 shows symbols represented by 3δ followed by δ and 4δ. The total time required to transmit these three symbols is 11δ rather

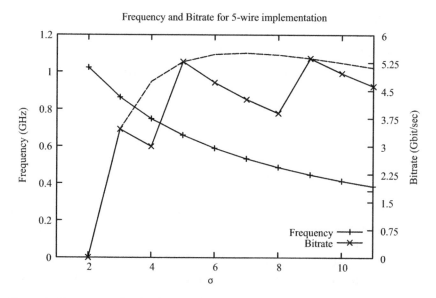

Figure 8.40   Bit rate for five-line time encoding.

than 15δ, and in the general case the average symbol time is reduced from 5δ to 3.5δ. The same argument can be applied to time encoding on multiple lines, where the symbol time needs only to be long enough to transmit the last transition of the symbol, but because there are many lines in which the last transition could occur, the saving in time is not so large. For a five-wire protocol with five time slots, there are 625 possible symbols, but 369 of them have a transition in the final time slot on one or more of the four stop lines. This means that the average symbol time can only be reduced from 6δ to 5.43δ.

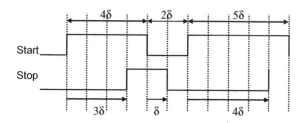

Figure 8.41   Variable symbol time.

8.6   PARALLEL SERIAL LINKS

As the number of lines in a parallel link is increased, the effects of skew, cross-talk, and process variability become more pronounced. To overcome these problems the time between symbols has to be increased, and so the link bandwidth does not increase proportional to the number of lines as might be expected. On the other hand a serial link does not suffer from skew because only one bit is sent at a time, and cross-talk from aggressor connections in the rest of the system is less well correlated with the data being transmitted, so there is little build-up of assisting or delaying effects. Fabrication processes below 65 nm show increasing amounts of variability which will affect the skew between lines, and the delays in the interconnect lines themselves increases substantially, so serial transmission has an advantage for nanometer processes, particularly because it is possible to use several serial links in parallel. In this situation the problem of correlated cross-talk still exists, but usually high-bandwidth serial links use differential signalling, so much of the cross-talk is cancelled out but at the cost of doubling the number of lines per serial link. Skew between serial links may also cause difficulties, but these can be overcome by buffering.

A de-skewing scheme using first in first out (FIFO) buffers is shown in Figure 8.42. Parallel data is written into the write register using a clock derived from the sender. Each bit in this register is sent along a serial link using a self-timed protocol such as LEDR, and then entered into a FIFO buffer at the receiver end of the link.

The time of arrival of each bit will be determined by the physical length of the serial link and variations in individual links. But provided

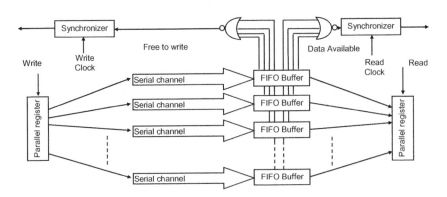

Figure 8.42   Skew buffering in serial parallel link.

there is sufficient space in the FIFO to accommodate the worst-case skew; all the bits from the sender will be stored as the first bit in the corresponding FIFO. Each FIFO has two flags, used for flow control of data. One indicates that the FIFO is empty, or that there are insufficient bits available in it to allow the read process to start. When all of these flags are low, there are sufficient parallel words available to start the read process. The other flag indicates that the FIFO is too full to allow any further writes. Only when all of these flags are low is the sender allowed to send more data.

The two flags are derived from the amount of data in the FIFO, and depend on both read and write processes, so if the read and write clocks are unrelated, it is necessary to synchronize the Free to Write signal to the write clock, and the Data Available signal to the read clock. Both the synchronization of the Free to Write signal and the serial transmission of data take a significant amount of time, and the sender cannot be stopped until this time has elapsed, so each FIFO must have sufficient space to contain all the data that could be transmitted between the Full flag going high and the last data bit arriving at the FIFO following the synchronized Free to Write signal going low. In practice this means that the Full flag goes high when there is still space to accommodate $(T_{synch} + T_{link}) f$ bits where f is the data rate in bits/second, T_{synch} is the synchronization time, and T_{link} is the link time delay. Similar considerations apply to the reader, because the synchronizer will delay the Data Available signal, so that the reader cannot be stopped immediately the FIFO becomes empty. This flag must then be set to allow sufficient data, $T_{synch} f$ bits, to accumulate in the FIFO before the read process is started. This type of scheme is discussed more fully in Section 7.2

A parallel serial system can produce a very high effective bandwidth, but may introduce considerable latency, because data cannot be read until after the buffers are sufficiently full. In order to achieve the potential bandwidth, the rate of reading and writing must be controlled so that the full or empty pointers are not set. For example, if a full pointer gets set, the write process will eventually be stopped, but there will be a long gap $(T_{synch} + T_{link})$ before it can be re-started, which leads to loss of effective bandwidth.

Much of the additional hardware would be required anyway in a high-bandwidth parallel transmission link, and synchronization and flow control are also needed in most data transmission systems, so the bandwidth advantages of using many serial links are likely to outweigh their disadvantages.

9

Pausible and Stoppable Clocks in GALS

Traditionally, clocks are derived from very accurate, crystal-controlled oscillators whose frequencies are fixed. Any interface between two independently clocked domains has to assume that the timing of the two clocks cannot be changed in any way, and this leads to the need for synchronizers to retime the data. In a globally asynchronous, locally synchronous (GALS) system [65,67,68], it is possible to pause, or stop the clock, rather than retime the data so that synchronization is not necessary.

There are two main methods of using the communication between processors to link the clocks. If the processor clock starts when data arrives, and activity stops as soon as the data is processed the clock is called data driven [71]. With a data-driven clock there is no choice between ongoing internal processor activity and the need to process new data, so there is no competition between them and there can be no metastable signals. A pausible clock [72,73] is normally running when the input data arrives so there is still a choice, to pause or not to pause, just before the next clock is due. Even if there is no internal processing when data arrives, requests coming from two independent sources to the same processor may have to be arbitrated. Instead of synchronizers with a fixed resolution time and a consequent probability of failure, a processor core with a pausible clock uses a MUTEX to decide if the clock should be paused in this cycle or the next. While the MUTEX is metastable, the clock is paused, the resolution time is unknown and can be unbounded, but there is no reliability problem. Pausing the clock

Synchronization and Arbitration in Digital Systems   D. Kinniment
© 2007 John Wiley & Sons, Ltd

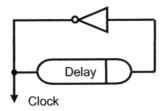

Figure 9.1   Delay-based oscillator.

means that the processor does nothing, but on average, the additional resolution time in the MUTEX is quite small. On average it is τ, so the loss of performance resulting from arbitration is also small.

9.1   GALS CLOCK GENERATORS

A simple oscillator made up from a delay and an inverter is shown in Figure 9.1, and this forms the basis of most stoppable clocks [74]. The delay itself is usually a chain of inverters and the total number of inverters in the ring is odd. This means the clock frequency is $1/2nt_d$ where n is the number of inverters, and t_d is the delay per inverter.

To ensure that the clock can only be stopped after a complete cycle we need a C gate [69,70] which only produces an output after all its inputs have changed. The top circuit in Figure 9.2 shows a simple design for a two-input C gate in which the output goes high after both inputs go high, and low when both inputs are low. If the inputs are different, the output will stay at its previous level. Circuits like this can also be modified so

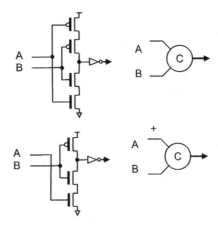

Figure 9.2   C Gates.

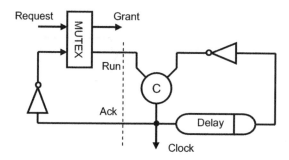

Figure 9.3 Pausible clock. Reproduced from Figure 1, "Demystifying Data-Driven and Pausible Clocking Schemes", by R Mullins and S Moore which appeared in Proc. Thirteenth Intl. Symp. on Advanced Research in Asynchronous Circuits and Systems (ASYNC), 2007 pp. 175–185 © 2007 IEEE.

that an input is only sensitive to a positive or negative edge. In the bottom circuit the output goes high only after both the A and B inputs have gone high, but only the B input needs to go low to give a low output, so A is sensitive to the positive going edge, but not the negative edge.

A complete pausible clock circuit is shown in Figure 9.3 [72–74]. The MUTEX outputs are both low if its inputs are both low, but if one input is high the corresponding output also goes high, leaving the other low. If both inputs go high, the MUTEX arbitrates between them. If the request was the first input to go high the grant signal goes high and run stays low, otherwise grant stays low and run goes high.

In Figure 9.3 with a low input request, the grant is always low and the run signal follows the inverse of the clock, so one input to the C gate changes when the inverse clock changes, and the other later at a time determined by the delay. As a result the clock cycles at a frequency of 1/2 Delay.

When the request signal goes high the run signal stays low if it is already low (clock is high), or goes low the next time the clock goes high. The clock cycle is completed by the other input to the C gate going low after the delay time. The clock then goes low, and a high-going edge arrives at the C gate after the delay. Because run remains low a new cycle starts. When the request goes low again, the clock is low, run goes high and a new cycle of the clock is initiated.

The clock circuit operates with a four-phase protocol:

- the request goes high, and the clock is paused;
- Ack goes low, signalling completion of the cycle;
- request is lowered, allowing the clock to continue;
- Ack goes high.

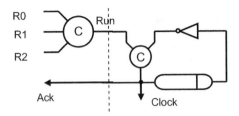

Figure 9.4 Synchronized inputs.

The alternative of a stoppable clock is shown in Figure 9.4 [71]. Here there are a number of request signals, R0, R1, and R2 indicating the arrival of a data item.

Run only goes high after all the data has arrived, and then Ack goes high. A single cycle of the clock is enforced by a four-phase protocol in which the requests are lowered, and Ack goes low after the end of the delay. A second cycle cannot start until new requests arrive and the stoppable clock is effectively synchronized to the data.

If two or more requests are competing for processor time, it is necessary to arbitrate between them. In Figure 9.5 the data-driven clock circuit is generalized to more than one input by replacing the C gate with an arbiter. Arbiters are discussed in more detail in Part III, here we assume that requests R0 and R1 arrive independently, and one is granted. The run signal then initiates a single clock cycle.

An arbiter can also be used as the input to a pausible clock as in Figure 9.6. In this circuit, as soon as the grant is made the request that succeeded is acknowledged. Pausible clocks which simply sample the request inputs every cycle, or wait for all requests to be present, are shown in Figure 9.7 and Figure 9.8.

In Figure 9.7 either R0 or R1 can pause the clock. The clock will not start again until the request is cleared, but in Figure 9.8 both R0 and R1 must be present before the clock is paused.

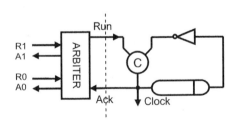

Figure 9.5 Arbitrated inputs. Reproduced from Figure 2, "Demystifying Data-Driven and Pausible Clocking Schemes", by R Mullins and S Moore which appeared in Proc. Thirteenth Intl. Symp. on Advanced Research in Asynchronous Circuits and Systems (ASYNC), 2007 pp. 175–185 © 2007 IEEE.

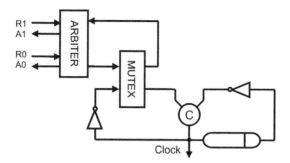

Figure 9.6 Pausible clock with arbitrated inputs. Reproduced from Figure 4, "Demystifying Data-Driven and Pausible Clocking Schemes", by R Mullins and S Moore which appeared in Proc. Thirteenth Intl. Symp. on Advanced Research in Asynchronous Circuits and Systems (ASYNC), 2007 pp. 175–185 © 2007 IEEE.

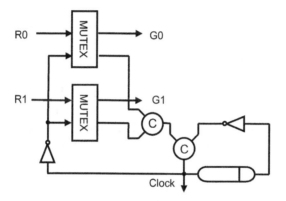

Figure 9.7 Pausible clock with sampled inputs. Reproduced from Figure 4, "Demystifying Data-Driven and Pausible Clocking Schemes", by R Mullins and S Moore which appeared in Proc. Thirteenth Intl. Symp. on Advanced Research in Asynchronous Circuits and Systems (ASYNC), 2007 pp. 175–185 © 2007 IEEE.

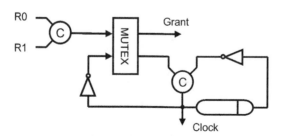

Figure 9.8 Pausible clock with synchronized inputs. Reproduced from Figure 4, "Demystifying Data-Driven and Pausible Clocking Schemes", by R Mullins and S Moore which appeared in Proc. Thirteenth Intl. Symp. on Advanced Research in Asynchronous Circuits and Systems (ASYNC), 2007 pp. 175–185 © 2007 IEEE.

As well as stopping or pausing the clock when an input request arrives, the clock may have to be to be stretched when output data has not been acknowledged, that is, the clock must remain low until output is complete.

9.2   CLOCK TREE DELAYS

The plausible and stoppable clock circuits described above rely on the clock tree delay being small. That means when the clock stops, the processor will stop in a relatively short time. In practice many synchronous processor designs are large and have a clock tree consisting of many stages to provide the drive necessary for a large system [75,76]. In a synchronous system the clock never stops so it does not matter that the delay through the clock tree is a large proportion of the clock cycle time provided that each edge arrives at the same time in all parts of the processor. If the clock has to be started and stopped by requests for a data transfer, all clock edges are associated with data, and it is important that the data and the clock edges that deal with that particular data item are timed to arrive at the same time.

Figure 9.9 shows how the insertion of a clock tree may affect the latching of data into the processor input register. In Figure 9.9 the clock tree is shown as a chain of buffers between the oscillator and the point where the input register is clocked. The clock is started when the run signal goes high, but there is a delay between the starting of the clock

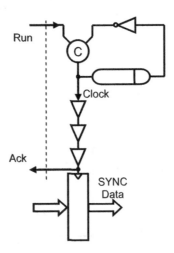

Figure 9.9   Effect of a clock tree delay. Reproduced from Figure 11, "Demystifying Data-Driven and Pausible Clocking Schemes", by R Mullins and S Moore which appeared in Proc. Thirteenth Intl. Symp. on Advanced Research in Asynchronous Circuits and Systems (ASYNC), 2007 pp. 175–185 © 2007 IEEE.

and its arrival at the input register. Because of this delay, the input request cannot be acknowledged until the data is safely clocked into the input register, thus delaying the data transfer by at least the amount of the clock tree delay. In many processors this delay is more than half the clock period, and it can be as much as several clock cycles in a large processor. Because the Run request cannot be removed until the acknowledge is sent, the clock cycle will be extended to at least twice the clock tree delay rather than twice the delay line time.

A simple solution to this problem for clock trees of less than one cycle time delay is shown in Figure 9.10 where the input data is clocked into a register by a clock unaffected by the tree delay, and then into the processor input register by a clock derived from the clock tree. The acknowledge can then be sent out much earlier, and the additional delay before the input data is usable by the processor may sometimes be used to do some pre-processing between the two registers. Effectively there are then two clock trees, the first is lightly loaded and drives the input register, and the second, loaded with the bulk of the processor clock burden, has a bigger delay of up to one clock cycle. The time between the data arriving in the first register and being clocked into the second may not be wasted, as it can be used to do some initial processing. Unfortunately this means that some redesign of the core processor may be necessary, which reduces the usefulness of a GALS system in reusing synchronous designs.

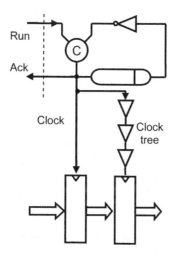

Figure 9.10 Offsetting the clock tree delay. Reproduced from Figure 11, "Demystifying Data-Driven and Pausible Clocking Schemes", by R Mullins and S Moore which appeared in Proc. Thirteenth Intl. Symp. on Advanced Research in Asynchronous Circuits and Systems (ASYNC), 2007 pp. 175–185 © 2007 IEEE.

If the clock tree delay is more than one cycle, it is necessary to break the request acknowledge handshake to the previous processor and buffer the data by inserting a FIFO in the data path. The number of registers in the FIFO must be at least equal to the number of rising clock edges in the clock tree at any time. New data waits in the FIFO for this number of cycles before being read into the processor, and there must be a guarantee that the delay between the entry of data into the FIFO and its arrival at the head of the FIFO is short enough to allow it to be read on the correct clock cycle.

This form of FIFO based buffering allows a high data rate, but introduces latency in the communication path at least equal to the clock tree insertion delay.

9.3   A GALS WRAPPER

In a stoppable or pausible clock GALS environment the synchronous core processors are embedded into an asynchronous communications framework by means of a wrapper which handles the input and output ports and the starting and stopping of the local clock [77]. An overview of such a wrapper is shown in Figure 9.11. FIFOs are used to buffer both the input and the output data, but for the FIFO a two-phase handshake is shown rather than a four-phase handshake protocol, so the arrival of new data at the input port is represented by a change in level on the input request signal, and the corresponding acknowledgement is represented by a change on the input acknowledge signal. The interface to the local clock is four-phase, with the availability of new data being signaled by a high level on the Request line, and acknowledged by the Grant signal.

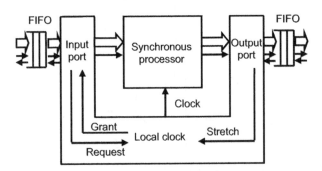

Figure 9.11   GALS wrapper. Reproduced from Figure 10, "Demystifying Data-Driven and Pausible Clocking Schemes", by R Mullins and S Moore which appeared in Proc. Thirteenth Intl. Symp. on Advanced Research in Asynchronous Circuits and Systems (ASYNC), 2007 pp. 175–185 © 2007 IEEE.

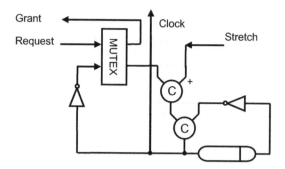

Figure 9.12 Local clock. Reproduced from Figure 10, "Demystifying Data-Driven and Pausible Clocking Schemes", by R Mullins and S Moore which appeared in Proc. Thirteenth Intl. Symp. on Advanced Research in Asynchronous Circuits and Systems (ASYNC), 2007 pp. 175–185 © 2007 IEEE.

Figure 9.12 shows the local clock circuit where a request pauses the clock. The clock restarts when the request signal is lowered, but when the synchronous core has new data to pass to the output FIFO, it may not be accepted immediately. In that case the output request and output acknowledge signals will have different levels, one high and the other low, and the output clock cycle must be stretched until such time that data is accepted. A stretch signal is produced which is low when the output handshake is incomplete, and a C gate with one input sensitive only to positive going edges is used to stretch the clock cycle.

The input port of the wrapper is shown in Figure 9.13. Valid data arriving at the head of the input FIFO generates a request, the clock is paused and the data is latched into the input register. At this point the request signal to the local clock is lowered, and the clock restarts,

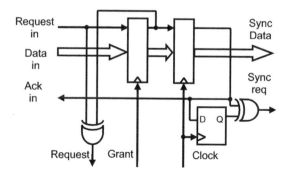

Figure 9.13 Input port. Reproduced from Figure 10, "Demystifying Data-Driven and Pausible Clocking Schemes", by R Mullins and S Moore which appeared in Proc. Thirteenth Intl. Symp. on Advanced Research in Asynchronous Circuits and Systems (ASYNC), 2007 pp. 175–185 © 2007 IEEE.

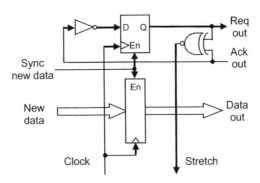

Figure 9.14 Output port. Reproduced from Figure 10, "Demystifying Data-Driven and Pausible Clocking Schemes", by R Mullins and S Moore which appeared in Proc. Thirteenth Intl. Symp. on Advanced Research in Asynchronous Circuits and Systems (ASYNC), 2007 pp. 175–185 © 2007 IEEE.

passing synchronized data to the core processor. On the following clock an acknowledge edge is sent to the FIFO ready for the next data transfer.

In the output port, Figure 9.14, valid new data is clocked into a register when it is available, and the request out set to the inverse of the acknowledge out. The stretch signal then goes low until the output FIFO has latched the data forcing the clock to stretch. When the data out is acknowledged the clock resumes.

10

Conclusions Part II

Systems on silicon are increasingly made of a number of independently timed core processors connected together by a communications network. Data passing between two of these processors needs to be retimed to allow it to pass between them safely, but the synchronization techniques that can be used depend on the timing relationship between the sender and the receiver.

In a synchronous system the timing of each process is usually driven by a single common clock. In a mesochronous system the clocks are more loosely linked by being phase locked to a common source so that phase can drift more widely, but within some bound. If the two clocks are effectively locked together in this way, the methods described in Section 7.3 avoid the need for conventional synchronization, and the latency of the interface can be small. In a plesiochronous system each processor may have its own autonomous clock, and Section 7.4 shows how data can be synchronized again, with relatively low latency if the clocks are similar in frequency, but the phase relationship is unbounded.

Latency and throughput are issues with more conventional synchronizers when the clock relationship between the two sides of an interface are unknown. In Section 7.1 a simple synchronizing interface is shown to have a relatively slow throughput, but throughput can be improved, usually at the expense of latency by adding a FIFO as in Section 7.2.

Relatively low-performance systems may not need a full clock cycle for synchronization, and that case part of the cycle can be used for processing as described by the LDL scheme of Section 7.5.1. On the other hand, nanometer fabrication processes may require synchronization times of more than one clock cycle. This is particularly true of high-performance systems where the clock period is only a small number of gate delays. The impact of this on latency can be severe, but reductions in latency are possible using techniques described in Section 7.5.2. Here, if a synchronization time of n cycles is needed to achieve adequate reliability, then it will usually be possible to approximately halve the synchronization time while accepting the possibility that a computation may have to be carried out again. In this case the latency is variable, but the worst case is no greater than that of the standard solution This re-computation procedure may take more than one clock cycle, but only need occur approximately once in several thousand data transmissions, so does not significantly impact on the system performance. Because the synchronization delay is reduced, data transmission latency is also reduced by a factor of approximately two when compared with more conventional synchronizers. Section 7.6 describes asynchronous communication mechanisms (ACMs) which are used in real-time systems where the data passed from one processor to another may not be held up by the reader or the writer, or both. They allow both processors to operate asynchronously by providing a common memory of three or for locations for the exchange of data. The concepts of coherency and freshness are introduced here and the algorithms described are informally shown to provide these properties.

The methods used to code and transmit information on a link between two processors are introduced in Chapter 8, and the advantages and disadvantages of different network architectures are discussed. The performance of a system depends to a large extent on its interconnectedness, and the bandwidth available to transmit information over the network of links. Various methods can be used, from simple serial links with NRZ coding to parallel methods using time endcoding. Their characteristics in terms of bandwidth, power dissipation, timing independence, and wire area are compared. In some cases several independent serial links can be used in parallel in order to increase bandwidth in places where there may be a bottleneck in a network on chip. In this case synchronization of the link is needed, but this will have some shared functionality with the existing need to synchronize data for the receiving processor.

An alternative to core processors with a continuously active clock is described in Chapter 9. If the clock itself is started by the arrival of data it does not need to be synchronized, or if it can be paused as soon as data arrives, conventional synchronization is not necessary either. Clock generators that are pausible or stoppable are described in Section 9.1, and a complete GALS wrapper for running synchronous core processors in an asynchronous environment is described in Section 9.3.

New synchronizers are often invented that claim to avoid the unreliability or unbounded delays implied by metastability, and some of them make it into print. Some of these traps for the unwary are given in Section 7.6, together with a description of how they are supposed to work, and why, in practice, they don't.

Part III

11

Arbitration

11.1 INTRODUCTION

Arbitration circuits, or simply arbiters, are commonplace in digital systems where a restricted number of resources are allocated to different user or client processes. Typical examples are systems with shared busses, multi-port memories, and packet routers, to name but a few. The problem of arbitration between multiple clients is particularly difficult to solve when it is posed in asynchronous context, where the time frames in which all clients operate may be different and vary with time. Here, the decision which resource is to be given to which process must be made not only on the basis of some data values, such as priority codes or 'geographical' positions of the clients, but also depending on the relative temporal order of the arrival of requests from the clients. In a truly asynchronous scenario, where no absolute clocking of events is possible, the relative timing order cannot be determined by any means other than making the arbiter itself sense the request signals from the clients at the arbiter's inputs. This implies the necessity to use devices such as synchronizers (Sync) and mutual exclusion (MUTEX) elements, described in previous chapters. Such devices have the required sensing ability because they operate at the analog level and they could be placed at the front end of the arbiter.

Why does the problem of arbitration need to be considered in asynchronous context? Can we build a clocked arbiter, in which the clock would tell the rest of the circuit when it should allocate the resources? If we attempt to answer these questions positively we will immediately run into an infinite implication. Namely: who will clock the arbiter?

Who will synchronize the clock with the rest of the arbiter? Who will clock the synchronizer (which is an arbiter itself!) between the clock and the rest of the arbiter? And so on.

The problem of designing systems with arbitration requires the designer to clearly understand the behaviour of typical arbiters in their interaction with clients and resources. Moreover, in order to design a new arbiter the designer should come up with a formal specification of the arbiter, by defining its architecture and algorithm, to be able to verify its correctness and possibly synthesize its logic implementation. This kind of task requires an appropriate description language, which should be capable of describing the various dynamic scenarios occurring in the overall system in a compact and functionally complete form. Unfortunately, traditional digital design notations such as timing diagrams, although adequate for simple sequential behaviours cannot easily capture asynchronous arbitration in a formally rigorous and precise way. Such behaviour can be inherently concurrent, at a very low signalling level, and may involve various kinds of choice, both deterministic and nondeterministic. In this chapter we are going to use Petri nets [78,79] and their special interpretation called signal transition graphs (STGs) [80,81] to describe the behaviour of arbiters. This modelling language satisfies the above-mentioned requirements. First attempts to use Petri nets and STGs in designing arbiters have been presented in [82,83].

Using Petri nets and STGs in describing arbiters enables the designer not only to trace the behaviour of the circuits, but also perform formal verification of their correctness, such as deadlock and conflict detection, as well as logic synthesis of their implementations. In this Part we do not consider these tasks focusing mostly on the architectural aspects of designing and using arbiters. However, we are currently working on a formal methodology for designing arbiters from Hardware Description Language specifications [84], where we are building on the initial ideas for synthesis presented in [85,86].

11.2   ARBITER DEFINITION

Let us start with a conceptual definition of arbiter. From a fairly general point of view, an arbiter has a set of input request channels that come from the clients which request resources, and it allocates resources by activating a set of output resource channels, each of which is associated with the provision of resources (Figure 11.1). The aim is to optimally match the resources requested with the resources available over a period of time.

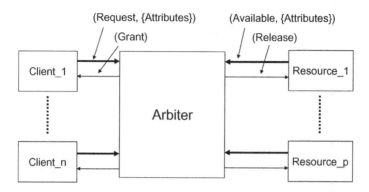

Figure 11.1 Synopsis of an arbiter.

The channels are subdivided further to signals, bearing in mind the asynchronous nature of protocols in which the arbiter interacts with its clients and resources. A particularly important role is given to handshake schemes, based on request/acknowledgement pairs of signals which will be defined in detail in the next section. For example, each request channel has a 'request' signal to the arbiter which indicates that resources are required and a 'grant' signal that is returned to the client when the resources requested are available. The request channel may also be equipped with some input data, 'request attributes', which indicate the priority of the request, and the type and amount of resources required (Figure 11.1). The data inputs are not allowed to change between request and grant.

Each resource channel has an input 'resource available' signal to indicate that the resource can be acquired by a client and an output 'release' signal which indicates the release of the resource by a client. The resource channel may have some input data, 'resource attributes', containing the characteristics of the resource available, such as for example the performance and capacity of the resource as some may be faster and larger than others (Figure 11.1). Again, the data must not change between resource available and release of the resource.

Therefore, informally, an arbiter is some sort of a dynamic 'matchmaker' between clients and resources which may follow some policy in making the matches. For example, in order to allow for fairness the arbiter may be required to be able to compute the number of requests granted from each input request and the number of output resource grants over a period of time. The arbiter can then ensure that the ratio between these counts is within certain limits for each client. How this is done is defined by the detailed behavioural description of the actual arbiter and its internal architecture, e.g. ring, tree, etc. At the algorithmic level, we may only allow one arbitration action between requests and resources active at one time, or we may have concurrency in the case of multiple resources so that groups of resources in a network could perhaps be allocated to competing clients while others are still active.

11.3   ARBITER APPLICATIONS, RESOURCE ALLOCATION POLICIES AND COMMON ARCHITECTURES

Examples of arbitration phenomena can be found at all levels of computing systems. In software for example, operating systems apply various mechanisms to synchronize processes in concurrent environments, the most well-known being E. Dijkstra's semaphores. A semaphore is a shared variable whose value can be tested and modified by several processes. Depending on the value the testing process may be admitted to a critical section of the code. Processes act as clients and critical sections act as resources. Semaphores, together with the software (or hardware) maintaining their test and modify operation, act as arbiters. On the other hand a semaphore variable itself, together with its operations, can be treated as a resource which needs protection in the form of another arbitration mechanism, implemented at the lower level of abstraction. At some level this recursive mutual exclusion problem has to be resolved by processor hardware. In the world of true concurrency where the processes are mutually asynchronous, e.g. some of them associated with peripheral devices, the problem is 'solved' by means of an interrupt mechanism that implements a set of binary signals, i.e. flip-flops, which record interrupt signals. There is a necessary element in processing interrupt signals, namely the method with which they are 'frozen' for processing. At the time when the interrupt signals are processed, their status and their data attributes must be stable to avoid hazards.

Having multiple layers of software deferring the ultimate conflict resolution down to the lower level is one possibility. However, in many cases, and particularly with developments in the area of systems-on-chip, one should seriously think about complex resource allocation procedures implemented directly in hardware. A typical example of arbitration at the hardware level would be a router or switch, say between processors and memory blocks, which has a number of input channels (clients) and output channels (resources) with a certain resource allocation mapping. For example, let the switch have three input channels, IC1, IC2 and IC3, and two output channels, OC1 and OC2. Let the allocation mapping be such that a request on IC1 requires use of OC1, IC2 on OC2, but a request on IC3 requires both OC1 and OC2.

In addition to resource allocation mapping, which does not say much about the fairness of resource allocation, there may also be fairness policies. A fairness policy may often be completely independent of the allocation mapping, but may also reflect the fact that certain mappings need more careful fairness policies. The issue of fairness is closely related with that of priority in the system. It arises when a number of requests can be served from the point of view of the allocation mapping, but the mutual exclusion conditions forbid granting them all. For example, in the above case, if only requests on IC1 and IC2 arrive, they can be served in parallel because there is no conflict between them on the required resources, OC1 and OC2, respectively. Similarly, if only an IC3 request arrives, it can be granted both OC1 and OC2. On the other hand, if all three requests arrive together, there may be an argument about unfairness to IC3. Indeed, because the requests may arrive not simultaneously, but with a small time difference, there is often a chance that the system may favour requests on IC1 and IC2 more than IC3. The reason for that is that the probability of the resources OC1 and OC2 being both free at the same time is much less than that of either of them being free, which implies the natural preference of the system towards IC1 and IC2, who need less resources, and may enter a mode where they alternate and leave IC3 starving. In order to affect the allocation policy, defined by the allocation mapping, one can introduce the notion of priorities which can exhibit themselves in different forms [87].

One typical form is a geographical position of the clients with respect to the arbiter which is built as a distributed system. The famous 'daisy-chain' arbiter, in which the requests of the clients are assumed to require only one resource are polled in a linear order, always starting from the same source position, has a fixed and linear priority discipline. Here the nearer the request to the source the higher its priority. Alternatively, one can apply a round-robin discipline in a ring-based architecture and launch one or more special 'tokens' associated with the resources. Each request can collect appropriate tokens, without releasing them until all of them are in possession so the grant can be issued. In this case, regardless of the relative position of the request, we can guarantee that after a finite number of cycles of circulating tokens all the tokens needed for the request have been collected and it will be granted.

Another form of introducing priorities is to attach them as data attributes to the requests, and involve them in the evaluation of the resource allocation mapping. However, in this case, there is still a possibility that several requests are in conflict. How to decide which of them should be granted? In other words, what sort of parameter should be added to the allocation algorithm in order to guarantee its determinism? The determinism can for example be ensured by having some history parameters stored with the arbiter. For example, in the above situation, we can store the fact that IC1 has been given OC1 k times in a row, and therefore it is not given OC1 again until IC3 has been granted.

Every application suggests a number of alternatives for the topological organization of a distributed arbiter, where ring, mesh and tree (cascaded) structures can be used. The combination of the structure and the resource allocation policy, including both its resource mapping and fairness disciplines, may affect the performance of the arbiter, for example its average response time. There is a trade-off between the average response time and fairness. A tree arbiter with a nearest-neighbour priority [88], for example, optimizes the average response time by passing the resource from a client to its nearest neighbour if the latter is requesting. This happens at the cost of being less fair to more distant neighbours. Another trade-off can be between the response time and energy consumption. For example, a ring-based arbiter, the so called 'busy ring', with constantly rotating token (which wastes power if there is no activity in the ring), has a better response than a ring arbiter with a static token, which stays with the most recent owner until requested by another request (and hence saves energy) [89].

While performance, power consumption and fairness are characteristics that can be quantified and traded off against each other, some properties are qualitative. These are mutual exclusion, deadlock-freeness, and liveness. They are general correctness criteria that the design must be verified for. For example, mutual exclusion can be defined for a concurrent system in the following way. Two actions are mutually exclusive if there is no state in which both actions are activated at the same time. A system is deadlock-free if there is no state in which no action is activated. A system is live with respect to an action if from every state the system can reach a state in which this action is activated. These properties can be defined formally using a rigorous notation, such as for example, Petri nets. The system can then be verified against these properties in terms of its Petri net model by various formal tools, such as reachability analysis or checking the unfolding prefix (see, e.g. [90,91]).

This section is organized as follows. We first look at the main hand-shaking protocols that are used for signalling between arbiters and the clients and resources. This is followed by the introduction of simple two-way arbiters based on synchronizers and MUTEXes. After that we consider more complex arbiters, first looking at the main types of multiway arbiters, built on cascaded (mesh and tree) and linear (ring) architectures. Finally, we look at arbiters with static and dynamic priorities. Here, we effectively treat them as event processors, whose role is not only to perform dynamic conflict resolution in time (saying which event happens first), but also carry out an important computational function associated with the allocation of a resource or resources. Such a separation of concerns is crucial for further development of both the architectural ideas about event processors as well as for the investigation of possible ways to automate their design process.

It should be noted that in this section we only consider arbiters with a single resource. This is not a limitation of our modelling or design approach, it is done mainly to keep the discussion sufficiently simple and concentrate on the aspects of the operation of arbiters from the point of view of its clients. Most of the architectures presented can be relatively easily applied to the case of multiple resources. The reader interested in arbitration problems with multiple resources can refer for example to [92], which describes the so-called Committee Problem. We recommend it as an exercise in applying Petri nets and STGs as a modelling language.

11.4   SIGNAL TRANSITION GRAPHS, OUR MAIN MODELLING LANGUAGE

In this section we briefly introduce STGs in order to equip the reader with the basic knowledge of the notation we shall use in this section to describe the behaviour of arbiters. We shall avoid here formal definitions and properties of STGs, as the purpose of using this model is purely illustrative. For a more systematic study of this model, including its use in synthesis and analysis of asynchronous logic with refer the reader to [107] and [108]. Informally, an STG is a Petri net [79] whose transitions are labelled with signal events. Such events are usually changes (i.e. rising and falling edges) of binary signals of the described circuit. An example of the STG which describes the behavior of a two-input Muller C-element is shown in Figure 11.2(a). Figure 11.2(b) represents a timing diagram description of the same behavior.

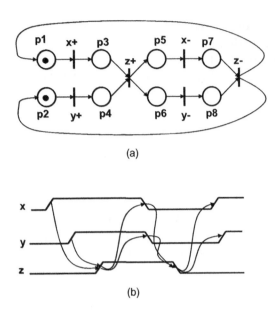

(a)

(b)

Figure 11.2   Modeling a Muller C-element: (a) signal transition graph; (b) timing diagram.

Here x and y are input signals, and z is the output of the circuit. The circuit changes its current output value (say, 0), when both its inputs transition from the previous value (0) to the new value (1). A similar synchronization takes place when the circuit's output is originally at logical 1. In the STG in Figure 11.2(a), signal events are labelled in the following way: $x+$ stands for the rising edge of signal x, and $x-$ denotes the falling edge.

The rules of action of the STGs are the same as those of Petri nets. Every time, all input places (circles in the graph, labelled p1, ..., p8) of some transition (a bar in the graph labelled with a particular signal event) contain at least one token (bold dot in the circle) each, the transition is assumed to be enabled and can fire. The firing of an enabled transition results in a change of the net marking (the current assignment of tokens to the places). Exactly one token is removed from each input place of the transition and exactly one token is added to each output place of the transition. The firing action is assumed to be atomic, i.e. it occurs as a single, indivisible event taking no time. An unbounded finite non-negative amount of time can elapse between successive firings (unless the transitions are annotated with delay information, e.g. minimum and maximum intervals, during which the transition can be enabled).

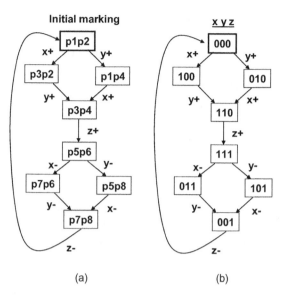

Figure 11.3 State graph for STG of Muller C-element: (a) reachable markings; (b) binary vectors.

The operation of the STG can be traced in the state graph, which is a graph of reachable markings obtained from the potential firings of the enabled transitions. For example, two or more transitions enabled in the same marking can produce alternative sequences of signals changes. If transitions are not sharing input places, like for example transitions $x+$ and $y+$ in Figure 11.2(a) ($x+$ has p1 as its input place, while $y+$ has p2), their firings can interleave, and form confluent 'diamond' structures in the state graph. The state graph for the above STG is shown in Figure 11.3, where part (a) depicts the reachable markings (lists of places marked with tokens), while part (b) shows binary vectors of signal values corresponding to the markings. The latter version is often called the binary encoded state graph of the STG.

For example, the initial marking of the STG (dotted places p1 and p2) corresponds to the top state of the graph. In that marking, all signals have just had a falling transition, so the state can be labelled by the vector 000 (with signal ordering xyz). Two transitions, $x+$ and $y+$, are concurrently enabled in that marking. Assuming $x+$ fires first, its successor (output) place p3 becomes marked, thus reaching the marking p3,p2, whose binary code is 100. Then $y+$ fires ($z+$ cannot fire yet, because only one of its input places is marked), thus reaching marking p3,p4, encoded as 110. Now $z+$ can fire, because both its predecessors are marked, and so on. The complete reachable state graph is constructed by exhaustive firing of the STG transitions, resolving concurrency in all possible ways (e.g. by firing $y+$ before $x+$ as well).

The state graph generated by an STG description of a circuit can be used for analysis and verification of the circuit's behaviour in its interaction with the environment. For instance one can state that the system modelled by an STG is free from deadlocks if every reachable state in its corresponding state graph has at least one enabled transition. Other properties can be derived, for example the fact that the model of an arbiter satisfies mutual exclusion can be found by checking if there is a state which can be reached by firing both grants (see Section 11.2). The state graph can also be used to derive the logical implementation of the circuit in the form of the functions of logic gates for all non-input signals. This requires some conditions to be satisfied, namely, that the transitions labelled with the same signal must interleave in the signs ("+" and "−") for any firing sequence, and that all the states that are labelled with the same vector have the same set of enabled non-input signal transitions. The details of the circuit synthesis method for STGs based on state graphs can be found in [107].

It should be noted that in the STG models we build for arbiters we often use the so-called short-hand STG notation, i.e. we omit places with only one input arc and one output arc, we place the labels of transitions directly, without showing a bar symbol. We will also use signal labels without a sign symbol, say x, to indicate any event on wire x, if we are not interested in the direction of the edge. This will be important for modelling arbiters in two-phase protocols.

12

Simple Two-way Arbiters

One of the goals of this chapter is to show that the MUTEX and Sync elements, studied in the previous chapters, can be used to build an arbiter of any complexity. For example, MUTEX and Sync are required in two different typical situations which are described in Sections 12.2 and 12.3. Section 12.1 first introduces the basic concepts and conventions used throughout the chapter: requests, grants, channels and protocols.

12.1 BASIC CONCEPTS AND CONVENTIONS

As described in Chapter 11, an arbiter is reactive to input requests coming from clients and produces output grants returned back to the clients. Because requests and grants are typically functionally tied into pairs, we use the concept of channel. A channel, which couples a request with a grant, constitutes a means of communication between the arbiter and a client in a bidirectional way (Figure 12.1). Similarly, communications between the arbiter and resources are organized using channels.

The communication behaviour of a channel can be defined using a protocol, called the handshaking protocol, which defines signalling and causality rules. Many different protocols exist. Only two of them are considered in this chapter because they are common and represent two general classes of protocols: those based on transition signalling, and those based on level signalling.

Synchronization and Arbitration in Digital Systems   D. Kinniment
© 2007 John Wiley & Sons, Ltd

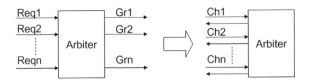

Figure 12.1 Coupling requests and grants into channels. Chn = (Reqn, Grn).

12.1.1   Two-phase or Non-return-to-zero (NRZ) Protocols

With this class of protocols, signalling is performed using signal transitions. In other words, only the occurrence of an event on a signal wire matters in this protocol, regardless of the initial and next level of the signal, hence its name non-return-to-zero (NRZ). Figure 12.2 describes a two-phase protocol for channel Ch1 as an example. Let us consider signals Req1 and Gr1 initially reset to zero (low level) by the client and the arbiter respectively. Then, a zero-to-one transition occurs on Req1 when client 1 is asking for arbitration. After processing the requests, the arbiter eventually issues a grant to client 1, by means of a zero-to-one transition on signal Gr1. This completes the two-phase protocol. The next communication will be using one-to-zero transitions (Figure 12.2).

It should be noted that the arbiter and the client must meet the following correctness conditions. Client 1 is not allowed to issue another request until the arbiter has responded with a grant. In the same way, the arbiter is not allowed to issue another grant until it has received a request.

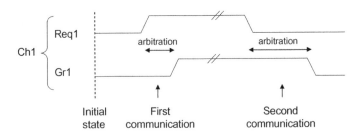

Figure 12.2 Two request/grant handshakes on channel Ch1 using a two-phase protocol.

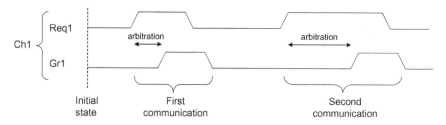

Figure 12.3 Two request/grant handshakes on channel Ch1 using a four-phase protocol.

12.1.2 Four-phase or Return-to-zero (RTZ) Protocols

This class of protocols uses level-based signalling as described in Figure 12.3. It is also known as a return-to-zero (RTZ) protocol. In the initial state, Req1 and Gr1 are both reset to zero by the client and the arbiter respectively. The communication starts when the client issues a request by setting Req1 to one. The arbiter then acknowledges the request by setting Gr1 to one. This completes the first two-phases of the protocol. Now starts the return-to-zero phase. In response to the setting of Gr1, the client resets Req1, and in response to Req1 the arbiter resets Gr1, which completes the four-phase protocol (back to the initial state).

It should again be stressed that for correct behaviour the client and arbiter must both respect the protocol rules. Req1 (Gr1) must remain stable at one or zero until Gr1 (Req1) has changed its state.

In the above protocols we assumed that the indication that the client has finished using the resource, i.e. the action that is often called the release of the resource is communicated on a separate channel, for example, the channel connecting the client directly with the resource. It is therefore not part of the arbitration procedure, which is a convenient way to separate concerns. However, sometimes people design arbiters in which the release action goes via the arbiter. For example, in the case of two-phase signalling, such an action uses an additional wire in the channel between the client and the arbiter (cf. signal Done in an RGD arbiter described in the next section). In the case of four-phase signaling, the release is often associated with the return-to-zero of the Req1 signal from the client. This happens often when a two-way or multi-way MUTEX is used as an arbiter itself. In the remainder of this chapter, we will assume the client to communicate the release action directly to the resource, unless it is specified otherwise.

12.2 SIMPLE ARBITRATION BETWEEN TWO ASYNCHRONOUS REQUESTS

Let us first revise the behaviour of a simple MUTEX element introduced in the previous chapters. The MUTEX, which can itself be used as a two-way arbiter, is described by the STG shown in Figure 12.4(a). The behaviour of this STG can be understood with the aid of the corresponding reachable state graph. In the STG, the initial marking contains a single token (i.e. single resource) in the place labelled me (mutual exclusion place). This token allows only one transition, either g1+ or g2+ (but not both!), to fire when both r1+ and r2+ have fired. The state graph clearly demonstrates that the arbiter model meets the requirement of mutual exclusion between the two grants, because although both transitions g1+ and g2+ are enabled in the state with the encoding 1100, only one of them can fire (disabling the other transition). A place such as me

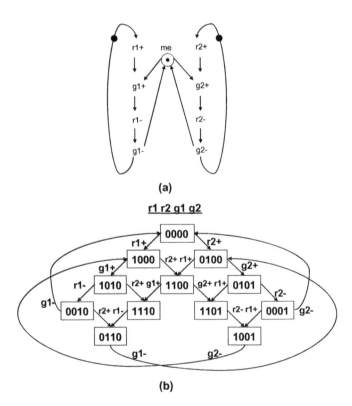

(a)

(b)

Figure 12.4 A two-way MUTEX (r1 and r2 are requests and g1 and g2 are grants): (a) STG; (b) corresponding reachable state graph.

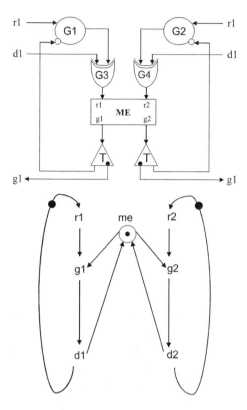

Figure 12.5   Request–Grant–Done (RGD) arbiter: (a) schematics; (b) STG specification.

is often called arbitrated choice (cf. controlled choice in the following examples).

The MUTEX uses the four-phase signalling on its request-grant pairs. All subsequent arbiter designs can be built on the basis of the MUTEX element. One such component, which is often itself a primitive element in many two-phase designs, is an RGD (Request–Grant–Done) arbiter, shown in Figure 12.5(a). Its input–output behaviour is described by the STG shown in Figure 12.5(b). Note that since this is a two-phase system the signal labels of transitions have no sign symbols.

Consider now a typical arbitration situation when choice is made between two clients, C1 and C2, who may request a unique common resource CR. This situation is represented in Figure 12.6, where the channel and signal names are defined. The arbiter must guarantee that one and only one client is accessing the resource, and that the handshake protocols are correct when a client accesses the common resource.

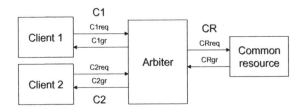

Figure 12.6 Basic architecture of a system enabling two clients C1 and C2 to safely access a common resource CR.

It is clear that clients C1 and C2 are in competition, via their communicating channels, when they both issue requests on signals C1req and C2req independently of each other. They have to be ordered in time in such a way that only one of them is chosen to be granted, thereby satisfying the property of mutual exclusion. Solving this problem requires a MUTEX element to decide between the two asynchronous requests and produce two mutually exclusive signals, C1req-arb and C2req-arb, to indicate the choice made. It is crucial that, once the decision has been made the MUTEX is not allowed to 'change its mind'. Thus, because C1req-arb and C2req-arb are mutually exclusive, they can safely be used to control the communication protocol. Figure 12.7(a) describes the arbiter's schematic for the four-phase protocol, in which gates G4 and G5 are Muller C-elements.

Figure 12.7(b) shows how a two-phase arbiter is built. The latter involves a relatively larger circuit, consisting of an RGD arbiter and a 2-by-1 Decision–Wait element [93] (also known as JOIN [94]). The RGD arbiter itself uses a MUTEX as a building block, as well as two toggles, two C-elements and two XOR gates. The behaviour of a Decision–Wait (DW) element is shown in Figure 12.7(c). It shows a nondeterministic choice (modelled by place p0) between events on inputs X1 and X2. Place p1 models controlled choice, i.e. only one of the two transitions z1 or z2 is enabled, due to the fact that either arc (x1, z1) will have a token or arc (x2,z2). When the DW element is used to implement a two-phase arbiter, shown in part (b), a nondeterministic (from the point of view of DW element) choice is resolved by the RGD arbiter, whose grant outputs are connected to inputs x1 and x2 of the DW element. Place p2 is an example of a so-called merge place, i.e. only one token arrives in it either after firing z1 or z2, but never both.

The composite behaviour of the two-phase arbiter solution is described by the STG shown in Figure 12.8.

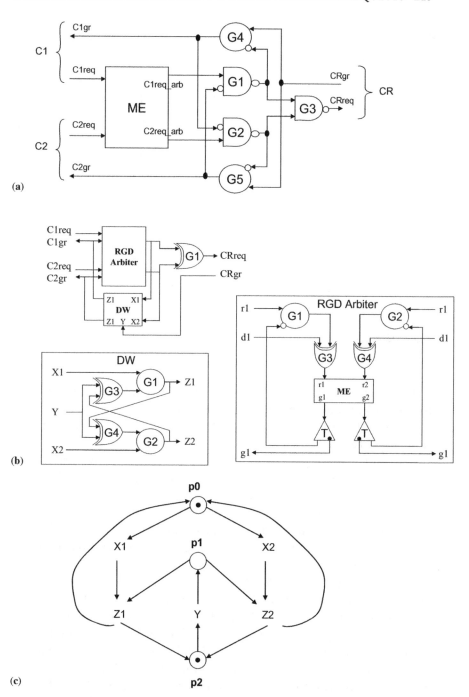

Figure 12.7 MUTEX-based arbiters: (a) for four-phase protocol; (b) for two-phase protocol; (c) STG for Decision–Wait.

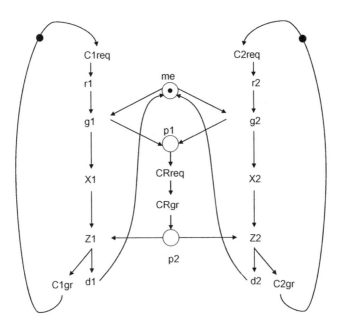

Figure 12.8   STG specification of the two-phase arbiter.

The behaviour of the four-phase one is shown in Figure 12.9. Let us, for example, examine the latter in more detail.

Assume that the request C1req from client C1 wins the arbitration. The MUTEX element then sets C1req-arb (transition g1+ fires) to one and maintains C2req-arb (i.e. grant g2) at zero until C1req is reset (transition r1- fires). Then, gate G1 and G3 fire and CRreq is set to one which issues a request to the common resource. Gate G4 is now enabled to fire as soon as CRgr is set to one. So, G4 fires and client C1 is granted (C1gr is set to one) which completes the two first phases of the protocol. Note that gate G2 is inhibited by C1gr (inverted input).

Now, C1req is reset by the client, C1req-arb (g1) is reset as well. G1 and G3 fire thereby resetting CRreq. At the same time, if C2req (r2) is one, the MUTEX may set C2req-arb (g2). G2 is still disabled, which postpones the request propagation through gate G2. CRgr is eventually reset and C1gr is reset, which on one hand completes the handshaking protocol on channel C1 and on the other hand opens G2 and the propagation of the request to the common resource. G2 and G3 fire, CRreq is set and another handshaking protocol starts on channel CR, which now communicates with channel C2.

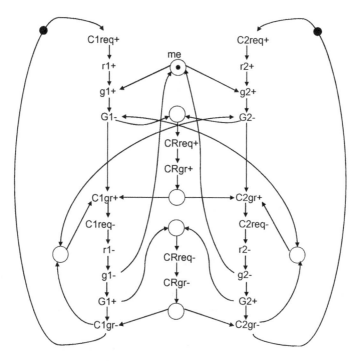

Figure 12.9   STG specification of the four-phase arbiter.

Finally, note that MUTEX is the key element which ensures mutual exclusion between the internal signals C1req-arb and C2req-arb. As long as this is guaranteed, the other gates maintain the proper signalling between the elected client channel and the resource channel, following the channel protocols.

12.3   SAMPLING THE LOGIC LEVEL OF AN ASYNCHRONOUS REQUEST

Another typical conflict resolution situation happens when there is a need to determine, or probe, whether a client issues a request or not. To do so, one needs to sample the logic value of the request signal which is asynchronous to the sample command. Synchronization is then required to determine if the level of the asynchronous request signal is zero or one. The reader may refer to previous chapters to explain the idea of a synchronizer in detail. For the purpose of carrying out synchronization, one could use one of those synchronizers. However, under certain

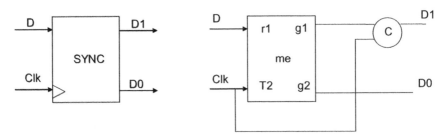

Figure 12.10 Sync, a synchronizer built around a MUTEX.

assumption about the protocol of interaction between the environment and the synchronizer, there is actually a possibility to build a synchronizer from a MUTEX element. This solution is shown in Figure 12.10.

The behaviour of this circuit, called Sync, is described by the STG shown in Figure 12.11. Basically, the circuit arbitrates between two rising edges of the data input D and the clock input Clk. If D wins, the circuit then waits for the arrival of the Clk before it responds with the rising edge on output D1, thereby indicating that the data is equal to one. If Clk wins, then the circuit responds with the setting of D0, thus saying that the data is equal to zero. The completion of the four-phase signalling in either case is based on the reset of either data input (for D1) or the clock input (for D0). This circuit is therefore similar to a

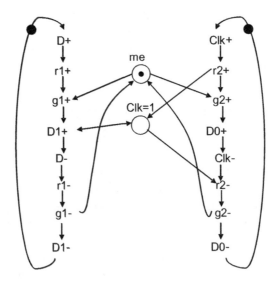

Figure 12.11 STG describing the behaviour of Sync.

MUTEX, with respect to following the four-phase handshake signaling on pairs (D,D1) and (Clk, D0) except that the former is enhanced in its setting phase by the condition Clk=1 when issuing the response on D1. Note that this STG uses the so-called read arc (dual-headed) which connects the Clk=1 place with transition D1+. The 'read-only' meaning of such an arc is in the fact that D1+ requires a token to be present in place Clk=1, but when D1+ fires this token is not consumed from this place. This place, however, has normal 'produce' and 'consume' relationships with transitions r2+ and r2, respectively.

Let us compare this circuit's behaviour with that of a more general type of synchronizer. This circuit, being speed independent (unlike the common synchronizer with a 'free-running' clock), assumes that the probing of signal D by the Clk input is only done when signal D is either stable in zero or in its rising mode, i.e. going from zero to one. Furthermore, it is assumed that this signal is only allowed to go back to zero after it has been eventually registered as a winner by the circuit. This is ensured by the four-phase handshake between D and D1. In other words, D is required to be persistent and monotonic.

Let us now consider a typical example of usage of a Sync element in an arbitration context. Consider a system which can be interrupted by a client named 'Interrupt Request'. The system normally performs task A in a loop, but every time the client 'Interrupt Request' is sending a request, the system inserts the execution of task B in the execution flow. Without loss of generality, assume that the execution of a task is performed by accessing the corresponding resource through a given channel. Then, the system's architecture can be described as in Figure 12.12.

The arbiter keeps running task A by communicating through channel A in a loop. Whenever 'Interrupt Request' sends a request through channel INT, the arbiter eventually responds to it by executing task B via the communication channel B.

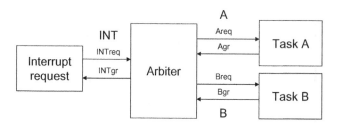

Figure 12.12 Basic architecture of a system with arbitration in the interrupt handling context.

To solve this problem we can again provide two kinds of solutions, one for the two-phase protocol and the other for four-phase. The STG describing the two-phase solution is shown in Figure 12.13, and the corresponding circuit in Figure 12.15(a). This solution uses an RGD arbiter to resolve arbitration between the INTreq and the polling signal, Agr, the grant part of the A channel. Normally, if there is no event on INTreq, every time Agr occurs, it wins arbitration via g1, which causes a response on Areq because an interrupt flag, var, would be in a zero state. If between two adjacent polling requests on A there is an outstanding event on INTreq, the latter would cause the RGD arbiter to issue a grant on g2, thereby causing the interrupt flag var to be set to one, after which the RGD arbiter is released by signal d2. Then, upon the next arrival of the polling event Agr, the g1 grant would trigger the alternative route in the Sel element, namely the one leading to Breq. As soon as an event Bgr is returned from Task B, it toggles the interrupt flag var back to zero, followed by the response INTgr and the production of the new polling event, acting as a proxy of Agr.

Observing the STG in Figure 12.13, we can notice the read arcs between place var0 (for the value of flag var=0) and transition Areq, and place

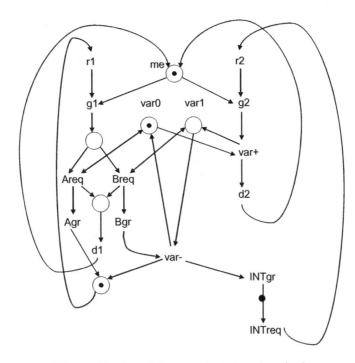

Figure 12.13   STG specification of the two-phase Sync-based arbiter.

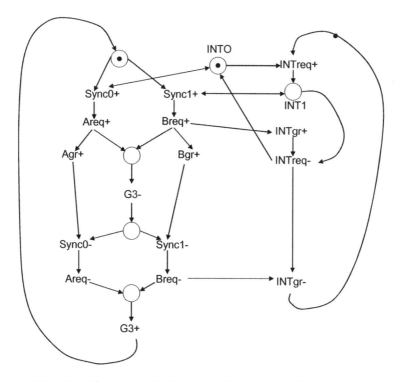

Figure 12.14   Specification of the four-phase Sync-based arbiter.

var1 and Breq. The token between these two mutually complementary places is toggled by transitions var+ and var−.

The four-phase solution is described by the STG shown in Figure 12.14, with the corresponding circuit in Figure 12.15(b). Here, because the system is based on levels rather than events, the arbiter has to check whether INTreq is high or not in order to determine if it has to perform a handshake on channel B or perform a handshake on channel A. A Sync block is therefore required to safely sample signal INTreq as shown in Figure 12.15(b). In order to represent the condition associated with the state of signal INTreq, the STG model again uses two complementary places INT0 and INT1 and read arcs controlling the enabling of transitions Sync0+ and Sync1+.

At reset, signals Bgr and Agr are zero and both Muller C-elements G1 and G2 are reset. Therefore, the output of the NOR gate G3 is set and INTreq is sampled by the synchronizer SYNC. Normally, if INTreq is low the SYNC block responds through its output 0, thereby causing a request on Areq through G2, followed by the reset of G3. This

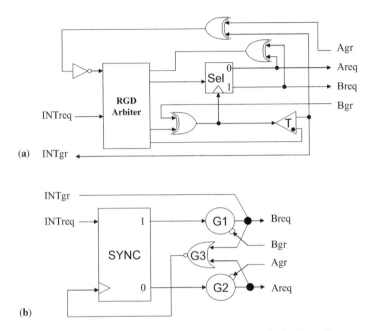

Figure 12.15 Sync-based arbiters: (a) two-phase protocol; (b) four-phase protocol.

is followed by the reset-to-zero phase on the A channel. Otherwise, if INTreq is high, the SYNC block replies through its output 1, it triggers G1 to perform handshakes on both INT and B channels. When they complete INTreq is sampled again, and so on.

12.4   SUMMARY OF TWO-WAY ARBITERS

In this section, we have introduced a channel as a means of providing a higher level of abstraction for communications between clients and resources via an arbiter. We have also introduced the two basic signalling schemes, two-phase (NRZ) and four-phase (RTZ), which can lead to different implementations and different interpretations of Petri net models as STGs. For the NRZ signaling the STG model is essentially event-based, i.e. the direction of the signal transition is not important. For the RTZ, the STG model has transitions in which plus and minus signs are meaningful. We have also defined two types of behaviour that are needed to model basic arbiters: making a choice between competing clients and determining if a client sends a request or not. Each of these behaviours requires the use of one of the two basic components, MUTEX and the Sync, respectively. Note that Sync can itself be efficiently built on the basis of a MUTEX, which leaves us with the important conjecture about the functional completeness of the two-way MUTEX (plus ordinary logic gates) in realizing any arbitration scheme.

The reader might have also noticed the significant difference in circuit complexity between NRZ and RTZ solutions, in which the RTZ solutions end up being more compact than the NRZ ones. This comes from the essential feature of the MUTEX being a level-sensitive element, due to the bias in its responses to rising and falling edges of request signals. As MUTEX is the basic building block for NRZ solutions, there is an inevitable area cost to be spent on the circuitry responsible for interfacing the MUTEX to the NRZ protocols between the arbiter and the clients.

In the following chapter we will show how more complex multi-way arbiters can be built using the two-way building blocks.

13

Multi-way Arbiters

The problem of multi-way arbitration has already been briefly discussed in Chapter 11. It is relatively easy to resolve arbitration in a two-way context by using a MUTEX element, built of a simple RS flip-flop and a metastability resolver. The latter is an analog, transistor-based, circuit which prevents the metastable state from its propagation from the outputs of the flip-flop to the outputs of the MUTEX. While the two-way MUTEX, built of CMOS transistors, is known to be free from oscillations, its generalization to the case of a three-way or n-way MUTEX, built on the basis of a multi-state flop, may actually exhibit oscillatory behavior due to the presence of the high-order solutions in its linear approximation [95,96]. As a result, it is recommended to build a multi-way MUTEX, and hence a multi-way arbiter with many clients, by means of composing two-way MUTEXes. Such compositions typically follow one of the standard topologies, such as a mesh, cascaded tree or ring. Other topologies, such as multi-ring and hybrid topologies, which are combination of the above-mentioned ones, are also possible. The detailed comparison of the arbiters within each particular topology and between topologies is beyond the scope of this chapter. Our goal here is to illustrate a number of designs, based on meshes, trees and rings, which all are built using the basic building block, a two-way MUTEX. We will also provide behavioural models for these solutions in the form of Petri nets. This should give the reader sufficient confidence that any other arbiter could be designed and modelled following a similar approach.

13.1 MULTI-WAY MUTEX USING A MESH

The idea of a mesh is quite simple. A two-way MUTEX is used for constructing arbitrating combinations on the 2-of-n basis. The example of a three-way MUTEX, consisting of three two-way MUTEXes is shown in Figure 13.1(a). The reader can easily check that this circuit correctly, in a speed-independent manner, implements its STG specification shown in Figure 13.2. Similarly, for $n = 4$ (5,...) one would need 6 (10,...) two-way MUTEXes, and so on. There could be many possible solutions for arbitrary n, depending on the requirements to the graph layout. For example, one possible layout could be obtained from the triangle under the main diagonal of the $n \times n$ matrixes shown in Figure 13.1(b). This layout is regular and avoids mutual crossing of interconnects between MUTEXes. It is easy to see that the complexity of the mesh architecture grows quadratically with the number of inputs n, which is certainly not economical. Morever, the latency, i.e. the propagation delay is proportional to n, which is also rather large.

This architecture is therefore not very popular if one needs to build an arbiter for sufficiently large n. However, for small values of n, such as

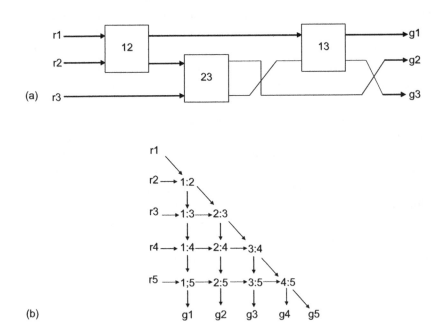

Figure 13.1 (a) Pair-wise mesh interconnection for three-way MUTEX; (b) idea for a regular layout.

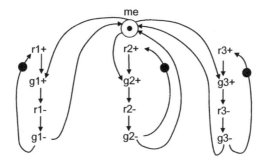

Figure 13.2 STG for a three-way MUTEX.

3 or even 4, this structure is quite practical, given that other topologies such as tree or ring require certain time and overhead due to the need for additional logic in the interface between the request–grant channels and the MUTEXes.

13.2 CASCADED TREE ARBITERS

Cascaded multi-way arbiters are based on a tree topology, in which the front end requests arbitrate in (adjacent) pairs, and then new requests are generated on their behalf. These new requests propagate to the next level of the tree and arbitrate with their neighbour in the same fashion, and so on. Finally, only two requests remain at the root of the tree. This structure is illustrated in Figure 13.3 for the case of $n=4$. In order to build such a cascaded arbiter for the four-phase (RTZ) signaling scheme, two

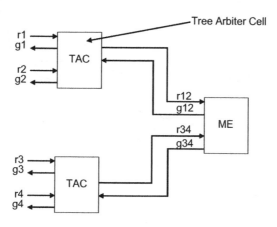

Figure 13.3 Four-way MUTEX built as a cascaded arbiter.

types of components are required. The basic one is a two-way arbiter, which was introduced in the previous section in Figure 12.6 and 12.7(a). We shall now call it a tree arbiter cell (TAC), in which the channel with the common resource is used to communicate with the higher level. For implementing the cell in the root of the tree, one can simply use a two-way MUTEX. This would give a multi-way MUTEX, where the request–grant protocols have the acquisition and release phase. For a multi-way arbiter that deals with a single resource, the root of the tree should be just another TAC. Here, the task of releasing the resource by the client would be implemented outside the arbiter, via a link between the client and the resource, exactly as it was assumed to be in Figure 12.6. One can build a two-phase (NRZ) version of the tree arbiter in a similar way using the two-way component of Figure 12.7(b).

The behaviour of a cascaded arbiter with NRZ signaling, acting as a multi-way RGD arbiter, where both acquisition and release phases propagate through the arbiter, is illustrated in Figure 13.4.

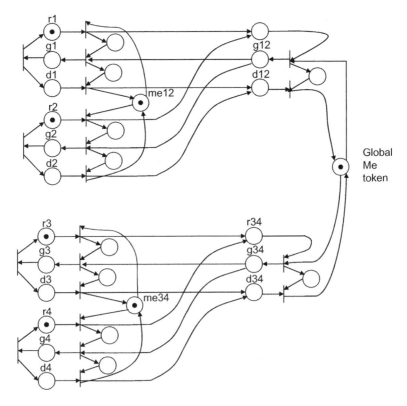

Figure 13.4 Petri net model of a cascaded RGD arbiter.

The analysis of the behaviour of the cascaded arbiter in either RTZ or NRZ form shows a number of important details that can offer a number of options in which the basic design of the arbiter can be modified.

First, the design can be modified by observing that the request produced in the TAC to the next stage (or to the resource in the case of two-way arbitration) does not need to wait until the mutual exclusion between the requests r1 and r2 has been resolved. Such arbiter designs, called low-latency or early-output arbiters, have been implemented in [82,97]. The idea behind such an arbiter can be seen in the STG shown in Figure 13.5.

Second, in the original tree arbiter design after the release of the request by the client, the release propagates all the way back to the root of the tree, regardless of the position where the next request is produced. A modification to that was proposed in a tree arbiter with the nearest-neighbour policy [88]. Here, at every TAC stage of the cascade, when a particular request is released, there is an additional checking of

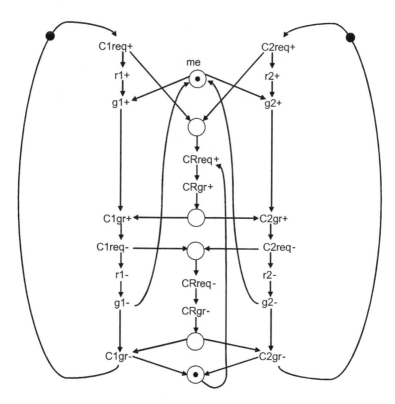

Figure 13.5 STG for a TAC with early output.

whether there is a request pending from the neighbour. If such request is pending the grant is 'short-circuited' at this level directly to the nearest neighbour. This modification, although affecting the fairness of arbitration by delaying the service of other neighbours, improves the throughput of the overall system. Intuitively, such a modification is quite tempting and seems to gain a lot in performance. However, in reality it does not! The detailed analysis of such an arbiter and its comparison with an ordinary (FIFO discipline) tree arbiter was carried out in [88]. This analysis shows that under realistic parameters of the system, such as delays in logic and interconnects and request and service rates (at exponential distribution), the gain in performance is normally only 20–30%, and certainly not a factor of two.

13.3   RING-BASED ARBITERS

Ring-based arbitration is usually based on the principle of a token ring. The multi-way arbiter, or better say the multi-way MUTEX because it has both the acquisition and release functions, is built of the nodes connected in a ring. Each node is connected to a client via request and grant lines (and a done line in case of the NRZ signalling for an RGD arbiter). There is no direct connection between the ring and the resource. The resource is represented by a token in the ring. This token can be acquired by the node if the client associated with the node wins the arbitration. Token ring arbitration can be organized in two ways, depending on the behaviour of the token at the time when there are no active requests from clients.

One approach, called a busy token ring or busy ring arbiter, is based on the idea of a token constantly rotating through the ring, visiting the nodes and polling the requests. The Petri net model of such an arbiter is shown in Figure 13.6. Here the place labeled 'Token arrived' enables the transition called 'Service' only when the Wait place has a token, which is the case when the Request is pending. As soon as the service is finished and the place ME is marked with the token, propagation of the privilege token continues.

Another approach, called the lazy ring, is based on the idea of keeping the token in the node where the client was the most recent winner. If some other client requests the token the information about this request will propagate to the location of the token and shift the token to the requesting node. Such movements should obviously involve arbitration at every intermediate node with their local requests. The Petri net model

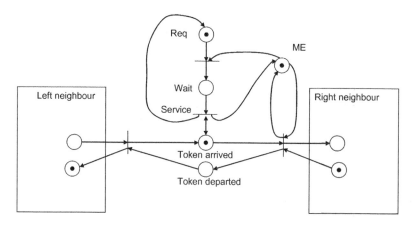

Figure 13.6 High-level model of a busy ring arbiter section.

of this behaviour is shown in Figure 13.7. Here, the position of the token, whether it is in this node or not, is indicated by the places called 'Token full' and 'Token Empty'. There are four possible scenarios, depending on the state of those places and depending on whether the local request or the request arriving from the left neighbour has won the arbitration in this stage. Consider for example, the situation when the 'Token empty' is marked and the winner is the left neighbour. This would cause the request to propagate to the right neighbour. Eventually the bottom transition on the right fires and switches the marking to the 'Token full' place. At this point the result of the local arbitration is used, i.e. the place 'Left neighbour won' is marked which causes the firing of

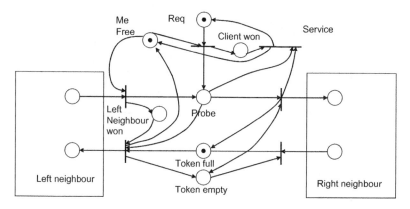

Figure 13.7 High-level model of lazy ring arbiter section.

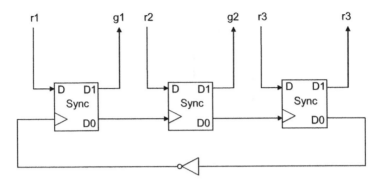

Figure 13.8 Three-way MUTEX implemented as a busy ring arbiter.

the bottom left transition. As a result the privilege token has moved to the left and the current stage has returned to the 'Token empty' state.

An example of possible implementation for the busy ring arbiter is shown in Figure 13.8. The main building block is a Sync synchronizer, which implements a node in the ring. In order to guarantee the appropriate

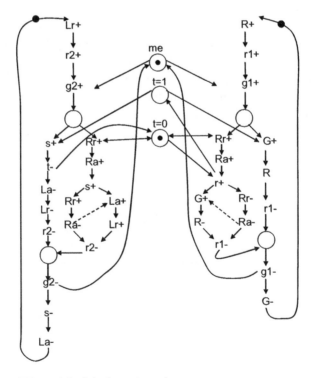

Figure 13.9 STG model of the lazy ring adaptor.

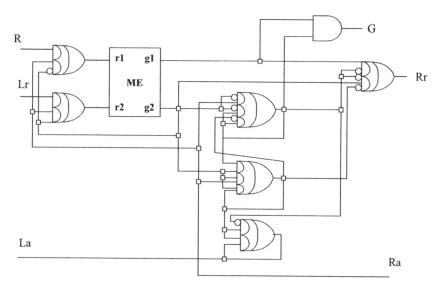

Figure 13.10 Circuit implementation of the lazy ring adaptor.

change of the phase of signal an inverter must be included. The RTZ signalling discipline is used here. The request polling action only happens on the rising edge of the token ring signal. The falling edge causes the global release of all the nodes. It could be possible to build a busy ring on the principle of an RTZ between the adjacent nodes, but this would require two handshake signals instead of one.

There have been several examples of implementing the lazy ring arbitration. For example, one was called distributed mutual exclusion [98]. Another possible implementation is shown here in the form of an STG for one node of a lazy ring arbiter (called a ring adaptor) in Figure 13.9. The circuit obtained from this STG is shown in Figure 13.10. This circuit was synthesized from this STG using methods described in [107].

The comparison of the busy and lazy ring arbiters has been performed in [89]. It showed that although the busy ring arbiter offers faster response time (its worst-case delay is proportional to the single ring delay while the lazy ring's worst case is twice as much), its excessive power consumption, when the traffic of requests is low, can be too prohibitive. Some hybrid solutions were proposed in [89].

14

Priority Arbiters

14.1 INTRODUCTION

In this chapter we present arbiters, originally presented in [87], which use priorities to generate grants. Traditional applications of arbiters with simple priority disciplines such as linear priorities and a priority ring are well known. Here, a more advanced formulation of a priority discipline is given, which is understood as grant calculation based on the current state of the request vector.

As an example of a practical application for such priority arbitration we recommend the reader to consider a fast network priority switch, shown in Figure 14.1. It has three input channels requesting access to a single output port. Every request is accompanied with a priority value transmitted through a dedicated priority bus. Priority values (attributes of requests) are generated dynamically. A dynamic priority arbiter takes a 'snapshot' of the request bus and calculates grant as a function of the request states (active or inactive) and the state of those priority busses which are accompanied with an active request. The priority discipline of an arbiter is formulated as a combinational function defined on the current state of request inputs, which is less restrictive than conventional, 'topological', mappings, such as that used in a daisy-chain arbiter.

Synchronization and Arbitration in Digital Systems   D. Kinniment
© 2007 John Wiley & Sons, Ltd

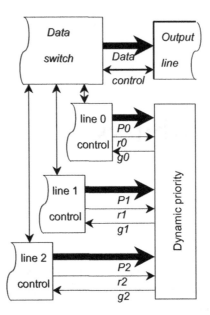

Figure 14.1 Network priority switch. Reproduced from Figure 1, "Priority Arbiters", by A Bystrov, D.J. Kinniment, and A.Yakovlev which appeared in ASYNC'00, pp. 128–137. IEEE CS Press, April 2000 © 2002 IEEE.

14.2   PRIORITY DISCIPLINE

All arbiters exhibit nondeterministic behaviour. The purpose of any asynchronous arbiter is to convert a stream of concurrent requests into a sequence of grants. Concurrency implies causal independence, which means that events on any two request inputs may occur in one order, in the opposite order or simultaneously. If two requests arrive with a significant separation in time while the arbiter is not busy with processing any other requests, then it is indeed impossible to reverse the order of grants by assigning different priorities to them. This creates confusion in understanding the concept of priority, which just does not work in the described scenario. All arbiters, with or without priority mechanisms, will behave similarly in this case.

However, arbiters differ in handling pending requests. Complex multi-way arbiters usually perform arbitration of pending requests concurrently with the execution of the critical section in the process which is currently granted. The arbitration results are stored in the internal memory and are used to compute a grant after the currently processed request is reset. Grant computation is illustrated in Figure 14.2. At first

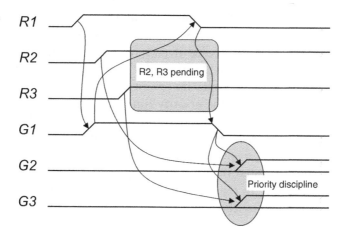

Figure 14.2 Priority discipline.

the three-way arbiter is idle, no requests R1...R3 are present and all grant outputs G1...G3 are in the passive state. Then the first request R1+ arrives. The first grant G1+ is generated immediately after R1+. There is no decision-making involved as R2 and R3 are still passive. Then two other requests arrive while G1 is still high. Propagation of these requests is blocked by G1, the effect known as request pending. Only after the first channel has finished and G1 switched to low, the other grants can be issued. The requests R2 and R3 are 'equal' at this stage, as both of them are active. The choice of which request propagates first can be either random or based on a rule of choice, otherwise known as a priority discipline.

The method of grant computation can be used for classification of complex arbiters. If such a computation takes into account the history of request processing (service history), then the arbiter is said to exhibit sequential behaviour or sequential priority discipline. Token ring, mesh and tree arbiters fall in this class. If the computation is entirely based on the current state of the request vector, then the arbiter has a combinational priority discipline. Such arbiters are referred to as priority arbiters (PA).

The following dimensions of the priority discipline taxonomy can be identified:

- memory (combinational–sequential behaviour);
- primary data (void–scalar requests–vector requests–request order);
- implementation (direct mapping–logic function).

These dimensions can be combined to describe a particular arbiter. For example, the tri-flop MUTEX, as in Figure 3.18, generates outputs according to the gate thresholds if two requests are pending. As no information is provided with the inputs for the conflict resolution, then the primary data type is 'void'. Its priority function is combinational and the implementation of the priority function does not belong to the logic domain (thresholds are analog values), which is typical for simple arbiters. Another example: the busy ring arbiter in Figure 13.8 belongs to the sequential type, its primary data is of scalar type and the priority discipline is implemented by the direct mapping of the model into the ring circuit structure. In the following sections we will describe several arbiters, which illustrate the above taxonomy.

14.3   DAISY-CHAIN ARBITER

The direct mapping implementation type of the priority discipline taxonomy is illustrated in the example of a daisy-chain arbiter with linear priorities (combinational priority function) and scalar requests.

The two-phase STG in Figure 14.3 shows that the priorities are defined by the order of request polling. Each stage either has a request outstanding at the time of polling if r is true, or not, if $\sim r$ is true. The order of polling, and hence the priority of each request cannot be changed without changing the topological structure of the system.

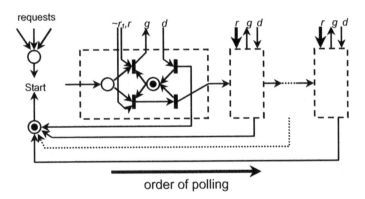

Figure 14.3   Daisy-chain arbiter model. Reproduced from Figure 2, "Priority Arbiters", by A Bystrov, D.J. Kinniment, and A.Yakovlev which appeared in ASYNC'00, pp. 128–137. IEEE CS Press, April 2000 © 2002 IEEE.

14.4   ORDERED ARBITER

The following design is an example of implementation of the request order type of the priority discipline taxonomy.

The ordered arbiter (OA), [104] uses the structure shown in Figure 14.4. It uses a FIFO to store grants in the order of request arrival and a request mask to minimize the amount of time when a request that is being granted is staying high, thus blocking the arbiter from its main function. This also maximizes the time when there are no requests pending on MUTEX inputs, thus improving its practical fairness and reducing metastability rate. The latency of such an arbiter in a busy system depends on the speed of FIFO, which can be made fast.

A three-way OA is shown in Figure 14.5. Interface and masking functions are performed by D-elements [102]. A request (r_1 for example) propagates to the n-input MUTEX input and further into the first channel of FIFO after arbitration. The left column of C-elements is a spacer, so the FIFO write acknowledgement is indicated at the second column as 0 on the output of the left NOR gate. This value is applied to the C-elements of the spacer, preparing them to accept an all-zero separation word. Only then a high level is set at the output of the upper AND gate

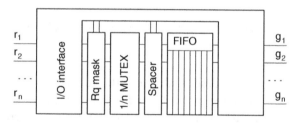

Figure 14.4   Ordered FIFO arbiter.

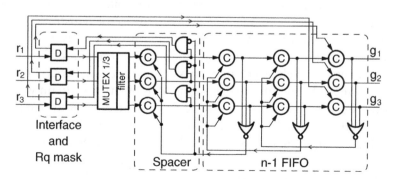

Figure 14.5   Ordered three-way arbiter circuit.

and applied to the D-element concluding the second phase of the four-phase handshake protocol. Then the D-element resets the MUTEX input to 0 (releasing MUTEX for the next arbitration), which propagates to the spacer and through the upper AND gate back to the D-element. At this point the value is written into FIFO and a separator word is formed. Then the D-element generates acknowledgement at its left output allowing the value to be shifted into the last FIFO stage. The critical section for the first client begins when g_1 is set high (it can be delayed by other values in FIFO). When the client drops its request, the value 0 is generated at the left output of the D-element. This value is applied to the C-element allowing a separator word into the rightmost stage of FIFO (reset g_1).

The time of the MUTEX being blocked is now bounded by a fixed delay $\tau = 2\tau_C + \tau_{NOR} + \tau_D$, which is small, whereas in an arbiter without a FIFO this delay is determined by the critical section of a client, whose bound is unknown.

One can see that this arbiter implements the simplest priority discipline based on the temporal order of requests. It can be modified to implement a different discipline, for example a LIFO, a FIFO with 'overtaking' or a disciple based on a combinational function using the global FIFO state as its primary inputs.

14.5   CANONICAL STRUCTURE OF PRIORITY ARBITERS

In the arbiter structure described below the function of registering the requests is completely separated from the combinational circuit implementing the priority function. Thus, a canonical structure is created which implements the combinational type of priority discipline taxonomy.

In order to be considered for the next grant, one or several requests must arrive before the moment of the request vector registration. This happens either if several requests arrive simultaneously (a low probability event) or if these arrive or stay pending during the critical section in one of the clients (a likely event in a busy system). The priority function can be changed by modifying the priority combinational circuit without altering the arbiter structure. Further, this can be done at initialization time by setting up additional inputs of the priority circuit, thus making the arbiter programmable. Such an arbiter can implement any priority function as opposed to an arbiter with a topologically fixed priority discipline, which is defined by the order of request polling.

Based on this concept, we construct a static priority arbiter (SPA), using single wires as request inputs, and a dynamic priority arbiter (DPA),

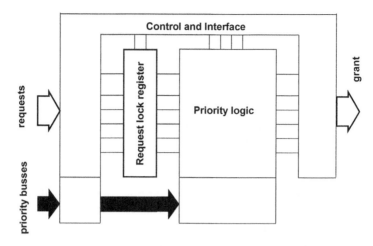

Figure 14.6 Static and dynamic priority arbiters. Reproduced from Figure 3, "Priority Arbiters", by A Bystrov, D.J. Kinniment, and A.Yakovlev which appeared in ASYNC'00, pp. 128–137. IEEE CS Press, April 2000 © 2002 IEEE.

whose request inputs are accompanied with buses carrying priority data. The DPA and SPA implement the scalar and vector primary data types of the priority discipline taxonomy correspondingly. In Figure 14.6 both SPA and DPA structures are shown.

SPA examples include an arbiter whose request lines form several groups, and the priority function always gives a grant to a member of the group with the greatest number of requests. Another example is context dependent arbitration, which gives a higher (or lower) priority to a request that belongs to the group of active requests forming a particular pattern.

A DPA uses dedicated priority buses to receive priority information from clients. The most obvious application of DPA is to detect a request with the highest priority value on the priority bus.

14.6 STATIC PRIORITY ARBITER

The structure of an SPA originates from the idea described in [104]. It operates in two stages. At the first stage it locks the request vector in a register comprising two-way arbiters (MUTEX elements). At the second stage it computes a grant using a combinational priority module. The separation of arbitration and grant computation functions allows the circuit to achieve truly combinational behaviour. Interface and control logic provides the correct functionality of the device under arbitrary gate delays. A single bit of the lock register is shown in Figure 14.7. It is similar to the request polling block of the token ring arbiter

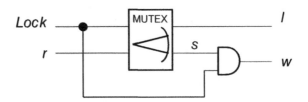

Figure 14.7 Lock register bit. Reproduced from Figure 5, "Priority Arbiters", by A Bystrov, D.J. Kinniment, and A.Yakovlev which appeared in ASYNC'00, pp. 128–137. IEEE CS Press, April 2000 © 2002 IEEE.

(see Section 13.3). The output of this circuit forms a dual-rail code, in which code words are separated by the all-zero spacer word. This facilitates the design of the priority module as a dual-rail circuit. Note, that for a speed-independent realization the AND gate in Figure 14.8 must be replaced by a C-element, or other appropriate means must be implemented to indicate the reset phase of the signals.

An example of a three-way SPA is shown in Figure 14.7. Let us examine its operation.

The input requests R1...R3 propagate through the set-dominant latches I1...I3, defined by the function $q^+ = set + q \overline{reset}$, to the MUTEX elements

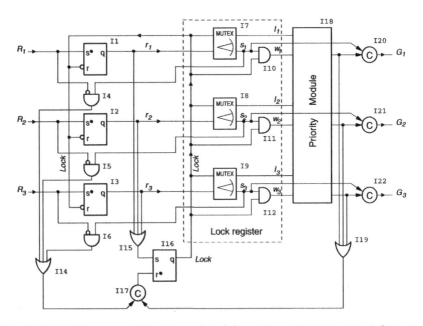

Figure 14.8 Three-way SPA. Reproduced from Figure 8, "Priority Arbiters", by A Bystrov, D.J. Kinniment, and A.Yakovlev which appeared in ASYNC'00, pp. 128–137. IEEE CS Press, April 2000 © 2002 IEEE.

I7...I9. After at least one of these reaches a MUTEX, a high signal level is formed at the output of the reset-dominant latch I16, defined by $q^+ = \overline{reset}.(set + q)$. This signal (Lock) is applied to the upper inputs of MUTEX elements, thus locking the request vector. After the input vector is locked, a dual-rail signal is formed at the inputs of the priority module. The priority module is implemented as dual-rail logic to guarantee that the output signal is generated after the process in the MUTEX elements is settled. The priority module determines which request should be granted and raises the signal at the appropriate output. This signal becomes locked in the output buffer comprising C-elements (I20...I22). After using the resource, the client resets its request. This is detected by I4...I6 and causes switching of I17 and I16, resulting in the resetting of the Lock signal. This resets the input latch, the MUTEX in the channel where the request has been dropped and all input signals of the priority module. Then the priority module outputs and the grant signal are reset. Then low levels are formed at the outputs of I14 and I19, which switches the C-element I17, unblocks I16 and enables processing of the next request. Correct operation of the lock register is crucial for the priority arbiter. An STG in Figure 14.9 shows how the polling cell is made insensitive to the gate delays. This STG describes the behaviour of a single channel, which explains why the indices after R, r, s, l, w and G signal identifiers corresponding to the request channels are not shown.

During its operation, the arbiter produces the following traces corresponding to the STG in Figure 14.9. If, at the moment of the Lock setting, the request R is not present, then the trace (Lock+, l+, Lock−/1, l−) is produced, which means that the polling cell has registered a 'lose'

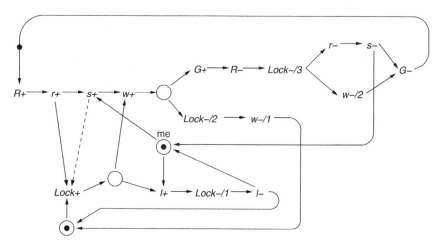

Figure 14.9 STG of the request lock register. Reproduced from Figure 7, "Priority Arbiters", by A Bystrov, D.J. Kinniment, and A.Yakovlev which appeared in ASYNC'00, pp. 128–137. IEEE CS Press, April 2000 © 2002 IEEE.

situation and the output grant has not been issued. If, at the moment of the Lock arrival, the request R is present, then two situations may take place. The first is described by the trace (R+, r+, s+, Lock+, loop:{w+, priority-resolution, Lock−/2, w− /1, Lock+}), which covers the scenario when the request is locked (s+), but the grant G is not issued. This happens if the priority module decides to satisfy a higher priority request in another channel that is not shown. The trace part labeled as 'loop' is executed repeatedly until the request R is satisfied. The second situation, in which the priority module satisfies the request R, is described by the trace (R+, r+, s+, Lock+, w+, priority-resolution, G+, R−, Lock−/3, r−, s−, G−). These traces clearly show that a dual-rail code is formed by the polling cell at its outputs w, and l, every time after a Lock+ event. The code words are separated by all zero words (w=0 and l=0), which facilitates the use of a dual-rail priority module.

The STG in Figure 14.9 indicates a potential problem in the realization of the request lock register. A scenario is possible in which the request R+ arrives, propagates to the MUTEX input causing r+ and activates s+ which, in this example, is considered to be extremely slow. At the same time r+ triggers Lock+ (dashed arrow) and subsequently l+ (both are fast). This creates a situation when a single request causes an all-lost vector being locked. The latter leads to a deadlock in the system as such a vector produces no priority module output, the C-element I17 (Figure 14.8) never switches to 1 and the Lock signal never resets. However in the practical realization of a two-way MUTEX based upon a cross-coupled latch this scenario is not possible due to the physics of the s+ transition. Once enabled, the voltage of an internal signal (which is the source of s) begins to change, which results in s+ rather than l+ even if Lock+ is very fast (1 ps in 0.6 μ technology). This very reasonable, in our opinion, timing assumption can be avoided (at the expense of the speed reduction) if the inputs of I15 Figure 14.8 are connected to the MUTEX outputs s1 ... s3. This will correspond to the dashed arc in Figure 14.9.

In the arbiter circuit shown in Figure 14.8, the only gate output which is not properly acknowledged is that of the OR gate I15. It can either be considered as part of a complex gate inside the I16 latch, or its delay must be bounded by a reasonable value. Both timing assumptions on the speed of the MUTEXes I7 ... I9 and the OR gate I15 can be removed if an additional priority module output is added which represents a 'nil grant'. This output would go high when an 'all-lost' request vector is locked (the priority module input labelled as 'not possible' in the truth tables below). The 'nil grant' can be used to reset the lock latch I16 (an

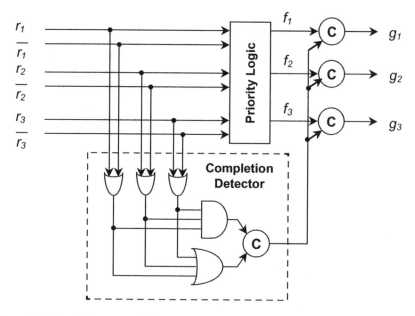

Figure 14.10 Priority module. Reproduced from Figure 8, "Priority Arbiters", by A Bystrov, D.J. Kinniment, and A.Yakovlev which appeared in ASYNC'00, pp. 128–137. IEEE CS Press, April 2000 © 2002 IEEE.

additional dominant reset input is needed) thus restarting the process of request locking and grant computation.

A possible dual-rail realization of the priority module is shown in Figure 14.10. The module produces 1-hot output after the dual-rail input is settled. The output becomes reset to an all-zero input word (separator). The hazard-free priority logic (PL) can be implemented as sum-of-products circuit without inverters. The truth table of this PL shown in Table 14.1 which implements a linear priority system.

Table 14.1 Linear priority function.

Input			Grant
$r1$	$r2$	$r3$	gi
l	l	l	Not possible
w	l	l	$g1$
l	w	l	$g2$
l	l	w	$g3$
w	w	l	$g1$
l	w	w	$g2$
w	l	w	$g3$
w	w	w	$g1$

Table 14.2   Priority function supporting groups.

Input				Grant
Group 1		Group 2		
r1	r2	r3	r4	gi
l	l	l	l	Not possible
w	l	l	l	g1
l	w	l	l	g2
l	l	w	l	g3
l	l	l	w	g4
w	w	l	l	g1
l	w	w	l	g2
l	l	w	w	g3
w	l	l	w	g1
w	l	w	l	g1
l	w	l	w	g2
w	w	w	l	g1
l	w	w	w	g3
w	l	w	w	g3
w	w	l	w	g1
w	w	w	w	g1

Symbol w in Table 14.1 means 'win' and signifies the request locked in the MUTEX register. Symbol l means 'lose' and indicates the request which is considered to be absent at the moment of arbitration.

A more complex example of the priority function is shown in Table 14.2. It handles requests that form two groups group1 = {r1; r2} and group2 = {r3; r4}. The higher priority is always given to a group with the largest number of active requests. If the number of requests is equal, then group1 wins. Within a group a request with the lower number has greater priority. This priority function can also be implemented as a sum-of-products hazard-free circuit.

14.7   DYNAMIC PRIORITY ARBITER

A dynamic priority arbiter (DPA) considered in this section performs comparison of the priorities supplied with each request and grants the one with the highest value. Note that the proposed structure may implement an arbitrary priority function, which is 'encoded' into the priority module. The structure of a DPA is very similar to the structure of an SPA as shown in Figure 14.6. The difference is that DPA uses additional data (priority values) produced by clients and passed into the DPA through dedicated priority busses. The value on a priority bus is valid only when the corresponding request is present. This creates a problem with speed-independent realization of the priority module as a delayed transition on the priority bus may be interpreted as an incorrect value. Another problem is to design a comparator for k n-bit priority buses which take into account only the values accompanied by a request. Disregarding the priority bus state is a nontrivial task in a speed-independent design, as special means are required to prevent hazards. In both the speed-independent and the bundled-data approaches a special signalling is needed to distinguish the data which have not arrived yet due to the data path delay from the data that will not arrive because the corresponding request is not set. The first problem is solved by using a dual-rail, or a 1-hot code, on the priority buses of the quasi-speed-independent arbiter realization. This is not needed for the bundled-data realization. The second problem is solved by introducing two signals v ('valid') and i ('invalid') which accompany every priority bus. These signals are generated by the lock register (similarly to w and l signals in the SPA) and propagated through the tree comparator distinguishing the priority value which has not reached the node yet from the absent request.

A 'practically acceptable' (see previous discussion about some minor delay assumptions) speed-independent DPA realization with 8 requests and 4 request priority values is shown in Figure 14.11. Priority buses

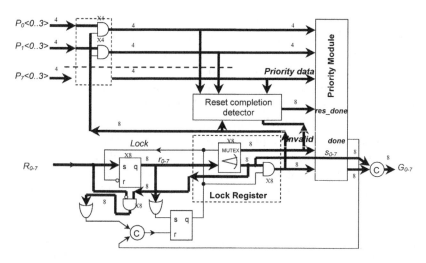

Figure 14.11 Dynamic priority arbiter. Reproduced from Figure 9, "Priority Arbiters", by A Bystrov, D.J. Kinniment, and A.Yakovlev which appeared in ASYNC'00, pp. 128–137. IEEE CS Press, April 2000 © 2002 IEEE.

carry a dual-rail code, though a design with 1-hot code has also been considered. The lines of the dual-rail priority buses $P_0 \langle 0{:}3 \rangle$ .. $P_7 \langle 0{:}3 \rangle$ are encoded as follows: $P_i \langle 0 \rangle$ is the lower bit, $P_i \langle 1 \rangle$ is the higher bit, $P_i \langle 2 \rangle$ and $P_i \langle 3 \rangle$ are complementary bits defined as $P_i \langle 2 \rangle = \overline{P_i \langle 0 \rangle}$ and $P_i \langle 3 \rangle = \overline{P_i \langle 1 \rangle}$. Gate arrays in this Figure are labelled as x4 or x8 depending on the number of array elements. If a bus is connected to such an array, then every bus line is assumed to be connected to the gate with the index number corresponding to the index number of the line. The lower part of the DPA schematic is similar to SPA. The upper part, comprising priority buses, AND gate arrays and a reset completion detector is new.

Several aspects enhancing functionality and improving performance were considered in the design of this arbiter. Arbitration and grant computation are separated, which permits realization of an arbitrary priority discipline. The control flow and the priority data flow are maximally decoupled in the priority module. This permits control signals to run forward without waiting for the slow data if the data are not needed to compute the result. For example, if a single request arrives, then the output of the priority module and the grant are generated almost immediately. Only the done signal waits for the completion of data manipulation, which happens concurrently with a critical section in the client process. Handshake phases in the control and data flow are also decoupled. C-elements are replaced by AND gates where possible and a reset phase of their inputs is indicated by additional 'reset-done' detectors.

The operation of the control part of the DPA in Figure 14.11 is similar to the SPA. The AND gates on the priority buses are only needed in the speed-independent realization to provide an all-zero separator word for the dual-rail priority module. The separator word (a.k.a. spacer) is indicated by the reset completion detector, which also checks valid-invalid signals for every channel. This detector comprises eight OR gates (one per channel). The inputs of every gate are connected to the lines of a priority bus and to the corresponding valid–invalid signals. Its output goes high as soon as the set phase in the channel begins (the completion of the set phase is indicated inside the priority module). The output goes low only after all inputs are reset establishing the separator word. The outputs of the reset completion detector are connected to the inputs of the reset completion system of the priority module. The priority module uses a tree structure as shown in Figure 14.12 to find the maximal priority value. Every node of the tree is a maximum calculation cell (MCC) which passes on the priority of the selected input, with the exception of the root node (RMCC) which only indicates completion.

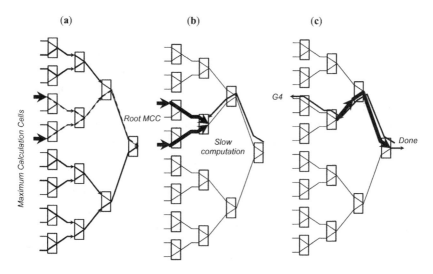

Figure 14.12 Accelerated grant generation: (a) fast control propagation; (b) grant propagation during data comparison; (c) early grant output. Reproduced from Figure 11, "Priority Arbiters", by A Bystrov, D.J. Kinniment, and A.Yakovlev which appeared in ASYNC'00, pp. 128–137. IEEE CS Press, April 2000 © 2002 IEEE.

Every cell has three modes of operation. If it gets 'invalid' signals in both channels, then the output 'invalid' signal is issued. All input priority data are disregarded and the output priority bus is set to the all-zero spacer word. If only one channel is 'valid', then its priority data are transferred to the output bus and the internal comparator is not invoked. If both channels are 'valid', then a comparator is used to compare the priority values and the greatest is propagated to the output. In the last two cases the output 'valid' signal is generated. The control 'valid' signal is permitted to propagate through the tree structure without waiting for the corresponding data. As a result these signals reach the root node very quickly and may even cause the RMCC to generate a primary grant. This happens if the RMCC receives a 'valid' signal in one channel and 'invalid' in the other. Further, such an accelerated grant may start propagating backwards if the nodes it propagates through do not need data to make a decision. Synchronization with datapath takes place only if the priority value is needed at the given MCC to decide to which MCC port the grant should be forwarded. This acceleration technique is illustrated in Figure 14.12. The nodes of the tree-type maximum calculation priority module are depicted as rectangles. Thick lines show propagation of the priority data. Thin dotted lines depict active 'valid' signals and continuous lines depict active 'invalid' signals. Arrows show the direction of signal propagation at the given stage. After 'valid' and 'invalid' control signals are produced by a request lock register, they begin to propagate through the maximum calculation priority module as shown in Figure 14.12(a). Their propagation is faster than the propagation of the priority data. A primary grant in Figure 14.12(b) is generated by the root node without waiting for data, because having one channel 'valid' and another 'invalid' is sufficient to decide which to grant. Then the primary grant begins reverse propagation to the nearest node which has both channels 'valid'. Only a node performing data comparison can delay grant propagation. After finishing the computation, the node permits grant propagation and at the same time transfers the maximum priority value to the output priority bus, as shown in Figure 14.12(c). This Figure illustrates that a grant can be released before priority data reaches the root node. This acceleration technique is similar to that used in a low-latency tree arbiter [82]. The difference is that in the described system slow events are associated with dataflow.

To minimize latency the dataflow and the control part ('valid'–'invalid' signals) are maximally decoupled. Synchronization is performed only when the data are needed to make decision. This results in better performance due to more concurrency. The accelerated propagation of control and grant signals does not indicate the set phase on the priority bus. Without this, one cannot guarantee a proper indication of the next phase, which is a reset. This problem is solved by using dual-rail (or 1-hot) circuits as comparators incorporated in MCC. As a result, a value on the priority bus at the root node becomes a code word (dual-rail or 1-hot) after the set phase on all 'valid' buses is completed. A completion detector in the root node generates a 'data done' signal, which is used to generate the output done. Additional signals 'reset done' indicate the reset phase on 'valid'–'invalid' lines and priority buses and allow replacement of C-elements by the faster AND gates in the node circuits. This also accelerates the resetting of the priority module because the control signals causing the reset are propagated without any synchronization at this stage (the cause is separated from the indication). The described delay-independent DPA with dual-rail priority busses has been implemented using an AMS 0.6 µ toolkit and Cadence CAD tools. After the design is initialized by applying the Reset signal, the request R0 is set. The arbiter replies with the grant signal G0 almost immediately (4.94 ns). It does not wait for the priority data $P0<0:3>$ to be set, as it is not required to generate the grant. The acceleration technique used in the priority module 'almost' disregards priority data when the output can be computed without it. An accelerated generation of the primary grant g0, can be done in 2.74 ns after V and I signals are set. The completion signal done is produced much later, when the propagation of data is finished. When many requests are pending and data comparison is performed in every level of the priority module the latency is increased to 7.63 ns.

Further speed improvements can be achieved by applying reasonable timing assumptions as shown in [87].

15

Conclusions Part III

In this Part we have considered a wide range of arbiters, from very simple two-way ones to multi-way ordinary, and then finally multi-way priority arbiters. We illustrated their behaviour by means of signal transition graphs, which are interpreted Petri nets.

Priority arbitration is a powerful method for resource allocation in the context of multiprocessor systems on chips. Here client processes gain access to a shared resource on a priority based discipline. A priority discipline is defined as an arbitrary combinational function, which is an extension to the traditional linear priority scheme. Typical applications for such arbitration schemes are (on-chip) busses and switches.

Conventional techniques for priority arbitration use disciplines which are fixed according to some topological order, such as, a daisy-chain. In the designs described above we allow the discipline of resource allocation to be a function of parameters of the active requests, which are assigned to the requests either statically or dynamically. We can thus consider systems which are clustered in terms of access to shared resources, and the resource allocation discipline is determined on the basis of a combinatorial function of a vector of active requests and their distribution among clusters. We illustrate this solution in our static priority arbiter. The static priority arbiter also provides a generic speed-independent circuitry to lock active requests and issue grants based on a priority module that can be 'plugged-in' or re-programmed with varying delays, without affecting the correctness of the overall logic. The dynamic case allows priorities to arrive together with their requests, and be involved in the process of computing the grant. In this dynamic case, we design a priority module which is built as a tree with speed independent priority data path encoding (dual-rail or one-hot). The special feature of this data path logic is that it uses a three-signal control ('valid'–'invalid'–'done'), which maximally decouples control flow from the data path by means of an early propagation of the 'valid'–'invalid' signals, concurrently with processing the priority data, separately acknowledged by 'done'. The data path computation can sometimes be 'disregarded' (depending on the 'valid'–'invalid' inputs) and grant issued with a very low latency. This acceleration significantly reduces the overall arbitration delay when the number of active requests is low. The design clearly illustrates the power of self-timed design principle in systems where delays are data dependent. We believe that the method with explicit 'valid'–'invalid' control can be extended to other applications involving data processing with asynchronously arriving data from multiple sources. One such application could be embedded microcontrollers.

References

[1] N. Rescher, Choice without preference: A study of the logic and the history of the problem of Buridan's Ass, *Kant-Studien*, 1959/60, No. 51, pp 142–175.

[2] S. Lubkin, Asynchronous signals in digital computers, *Discussion, Proc ACM*, 1952, pp 238–241.

[3] I. Catt, Time loss through gating of asynchronous logic pulses, *IEE Trans.*, 1966, EC15, 108–111.

[4] T.J. Chaney and C.E. Molnar, Anomalous behavior of synchronizer and arbiter circuits, *IEEE Transactions on Computers*, C-22(4), 412–422, April 1973.

[5] D.J. Kinniment and D.B.G. Edwards, Circuit Technology in a large computer System, *Proc. conference on Computers-Systems and Technology* London, October 1972, pp 441–449.

[6] D.J. Kinniment and J.V. Woods, Synchronization and arbitration circuits in digital systems, *Proc. IEE* 123(10), 961–966, October 1976.

[7] H.J.M. Veendrick, The behavior of flip-flops used as synchronizers and prediction of their failure rate, *IEEE Journal of Solid-State Circuits*, SC-15(2), 169–176, April 1980.

[8] K.O. Jeppson, Comments on the Metastable Behavior of Mismatched CMOS Latches, *IEEE Journal of Solid State Circuits* 31(2) 275–277, February 1996.

[9] C.L. Seitz, Ideas about arbiters, *Lambda*, 1 (first quarter):10–14, 1980.

[10] C. Dike and E. Burton, Miller and Noise Effects in a Synchronizing Flip-Flop *IEEE Journal of Solid State Circuits*, 34(6), 849–855, June 1999.

[11] N.H.E. Weste, and K Eshraghian, *Principles of CMOS VLSI design: A systems perspective*, 2nd edn, Addison-Wesley, 1992, p 326.

[12] F.U. Rosenberger, C.E. Molnar, T.J. Chaney and T-P. Fang, Q-Modules: Internally Clocked Delay-Insensitive Modules, *IEEE Transactions on Computers*, 37(9), 1005–1018, September 1988.

[13] C.H. van Berkel and C.E. Molnar, Beware the 3-Way Arbiter, *IEEE Journal of Solid-State Circuits*, 34, 840–848, 1999.

[14] O. Maevsky, D.J. Kinniment, A. Yakovlev, and A. Bystrov, Analysis of the oscillation problem in tri-flops, *Proc. ISCAS'02*, Scottsdale, Arizona, May 2002, IEEE, vol. I, pp 381–384.

[15] A. van der Ziel, Thermal Noise in Field Effect Transistors, *Proc. IEEE*, August 1962, 1801–12.

[16] G.R. Couranz, and D.F. Wann, The theoretical and experimental behaviour of synchronizers operating in the metastable region, *IEEE Transactions on Computers* C-24(6), 604–616 June 1975.

[17] S. Yang and M. Greenstreet, Computing Synchronizer Failure Probabilities, *Proc. DATE 07*, April 2007.

[18] D.J. Kinniment, A. Bystrov, A.V. Yakovlev, Synchronization Circuit Performance, *IEEE Journal of Solid-State Circuits*, 37(2), 202–209, 2002.

[19] C. Dike and E. Burton, Miller and Noise Effects in a Synchronizing Flip-Flop, *IEEE Journal of Solid State Circuits*, 34(6), 849–855, June 1999.

[20] QuickLogic Corporation, Metastability report for FPGAs, Application Note, 1997.

[21] D.J. Kinniment and J.V. Woods, Synchronization and arbitration circuits in digital systems, *Proc. IEE*, 123(10), 961–966, October 1976.

[22] M. Maymandi-Nejad and M. Sachdev, A digitally programmable delay element: Design and analysis, *IEEE Transactions on Very Large Scale Integration (VLSI) Systems*, 12(10), 1126–1126, October 2004.

[23] J. Zhou, D.J. Kinniment, G. Russell, and A. Yakovlev, A Robust Synchronizer Circuit, *Proc. ISVLSI'06*, pp 442–443, March 2006.

[24] K.A. Bowman, X. Tang, J.C. Eble, and J.D. Meindl, Impact of extrinsic and intrinsic parameter fluctuations on CMOS circuit performance, *IEEE J. Solid-State Circuits*, 35, 1186–1193, August 2000.

[25] W.J. Dally, and J.W. Poulton, *Digital Systems Engineering*, Cambridge University Press, 1998.

[26] T. Chelcea and S.M. Nowick, Robust interfaces for mixed-timing systems with application to latency insensitive protocols, *Proc. 38th ACM/IEEE Design Automation Conference*, pp 21–26, June 2001.

[27] A. Iyer and D. Marculescu, Power-performance evaluation of globally asynchronous, locally synchronous processors. *Proc. 29th International Symposium on Computer Architecture*, pp 158–168, June 2002.

[28] A. Chakraborty. and M. Greenstreet, Efficient Self-Timed Interfaces for crossing Clock Domains. *Proc. ASYNC 2003*, Vancouver, 12–16 May 2003, pp 78–88.

[29] A. Chakraborty and M.R. Greenstreet, A minimalist source-synchronous interface. *Proc. 15th IEEE ASIC/SOC Conference*, pp 443–447, September 2002.

[30] L.R. Dennison, W.J. Dally, and D. Xanthopoulos, Low latency plesiochronous data retiming. *Proc. 16th Anniversary Conference on Advanced Research in VLSI*, pp 304–315, 1995.

[31] M.R. Greenstreet, STARI: A Technique for High-Bandwidth Communication. *PhD Thesis*, Department of Computer Science, Princeton University, January 1993.

[32] M.R. Greenstreet, Implementing a STARI chip. *Proc. 1995 International Conference on Computer Design*, pp 38–43, Austin, Texas, October 1995.

[33] I. Sutherland and S. Fairbanks, GasP: A minimal FIFO control. *Proc. 7th International Symposium on Advanced Research in Asynchronous Circuits and Systems*, pp 46–53, April 2001.

[34] Y. Semiat and R. Ginosar, Timing Measurements of Synchronization Circuits. *Proc. ASYNC2003*, Vancouver, 12–16 May 2003, pp 68–77.

[35] R. Ginosar and R. Kol, Adaptive Synchronization. *Proc. AINT2000*, Delft, 19–20 July 2000, pp 93–101.

[36] R. Dobkin, R. Ginosar and C. Sotiriou, Data Synchronization Issues in GALS SoCs. *Proc ASYNC 2004* pp 170–179.

[37] R. Dobkin, R. Ginosar, and C. Sotirou, High Rate Data Synchronization in GALS SoCs. *IEEE Trans. VLSI systems* 14(1), 1063–1074, October 2006.

[38] N.A. Kurd, J.S. Barkatullah, R.O. Dizon, T.D. Fletcher, and P.D. Madland, Multi-GHz Clocking Schemes for Intel Pentium 4 Microprocessors, *Proc. ISSCC 2001*, February 2001, pp 404–405.

[39] S. Tam, S. Rusu, U.N. Desai, R. Kim, J. Zhang, and I. Young, Clock Generation and Distribution for the first IA-64 Microprocessor. *IEEE JSSC* 35(11), 1545–1552, November 2000.

[40] G.K. Konstadinidis, K. Normoyle, *et al.* Implementation of a third-generation 1.1-GHz 64-bit microprocessor. *IEEE Journal of Solid-State Circuits*, 37(11), 1461–1469, Novemer 2002.

[41] A. Kowalczyk, V. Adler, *et al.* The first MAJC microprocessor: a dual CPU system-on-a-chip. *IEEE Journal of Solid-State Circuits*, 37(11), 1461–1469, November 2001.

[42] P.J. Restle, T.G. McNamara, *et al.* A clock distribution network for microprocessors. *IEEE Journal of Solid State Circuits*, 36(5), 792–799, May 2001.

[43] F.U. Rosenberger, C.E. Molnar, T.J. Chaney, and T.-P. Fang, Q-Modules: Internally Clocked Delay-Insensitive Modules. *IEEE Transactions on Computers*, 37(9), 1005–1018, September 1988.

[44] A. Bystrov, D. Sokolov, and A. Yakovlev, Low Latency Control Structures with Slack, *Proc. ASYNC2003*, Vancouver, 12–16 May 2003, pp 164–173.

[45] H.R. Simpson, Four-slot fully asynchronous communication mechanism, *IEE Proceedings* 137(E1), 17–30, January 1990.

[46] H.R. Simpson, Correctness analysis of class of asynchronous communication mechanisms. *IEE Proceedings* 139(E1) 35–49, January 1992.

[47] F. Xia, A. Yakovlev, D. Shang, A. Bystrov, A. Koelmans, D.J. Kinniment, Asynchronous Communication Mechanisms Using Self-timed Circuits. *Proc. 6th Int. Symp. on Advanced Research in Asynchronous Circuits and Systems (Async2000)*, April 2000, Eilat Israel IEEE Computer Society Press, pp 150–159.

[48] R. Ginosar, Fourteen ways to fool your synchronizer. *Proc. ASYNC2003*, Vancouver, 12–16 May 2003, pp 89–196.

[49] A. El-Amawy, M. Naraghi-pour, and M. Hegde. Noise modeling effects in redundant synchronizers. *IEEE Trans Computers* 42(12), 1487–1494, December 1993.

[50] B.S. Landman and R.L. Russo, On a Pin Versus Block Relationship For Partitions of Logic Graphs. *IEEE Trans. Comput.*, C-20, 1469–1479, 1971.

[51] W.E. Donath, Placement and Average Interconnection Lengths of Computer Logic, *IEEE Trans. Circuits Syst.* CAS-26, 272–277, 1979.

[52] L. Benini and G. De Micheli, Networks on chip: a new paradigm for systems on chip design *Design, Automation and Test in Europe Conference and Exhibition, Proceedings DATE* 2002. pp 418–419.

[53] E. Bolotin, I. Cidon, R. Ginosar, and A. Kolodny, QNoC: QoS Architecture and Design Process for Network on Chip, *Journal of Systems Architecture*, special issue on Network on Chip, 50, 105–128, February 2004.

[54] E. Bolotin, I. Cidon, R. Ginosar and A. Kolodny, Cost Considerations in Network on Chip. *Integration-The VLSI Journal*, special issue on Network on Chip, 38(1), 19–42, October 2004.

[55] *International Technology Roadmap for Semiconductors (ITRS)*, Semiconductor Industry Association, 2001.

[56] T Bjerregaard and J Sparso, A Scheduling Discipline for Latency and Bandwidth Guarantees in Asynchronous Network-on-Chip. *Proc. 11th IEEE International*

Symposium on Asynchronous Circuits and Systems, Salt lake City 2005, pp 34–43.

[57] R. Ho, K.W. Mai, and M.A. Horowitz, The Future of Wires, *Proc. IEEE*, 89(4), 490–504, 2001.

[58] R. Dobkin, Y. Perelman, T. Liran, R. Ginosar, and A. Kolodny, High Rate Wave-Pipelined Asynchronous On-Chip BitSerial Data Channel. *Proc. 13th IEEE International Symposium on Asynchronous Circuits and Systems*, Berkeley, CA 2007, pp 3–14.

[59] M.T. Dean, T. Williams, *et al.* Efficient Self-timing with Level Endcoded 2-Phase Dual-Rail (LEDR). *Proc. ARVLSI*, pp 55–70, 1991.

[60] W.J. Bainbridge and S.B. Furber, CHAIN: A Delay Insensitive CHip Area Interconnect. *IEEE Micro special issue on Design and Test of System on Chip*, 142(4), 16–23, September 2002.

[61] C. D'Alessandro, D. Shang, A. Bystrov, A. Yakovlev, and O. Maevsky, Multiple-Rail Phase-Encoding for NoC. *Proc. 12th ASYNC*, pp 107–116, March 2006.

[62] E. Raisanen-Ruotsalainen, T. Rahkonen, and J. Kostamovaara, Time interval measurements using time-to-voltage conversion with built-in dual-slope A/D conversion. *Proc. 1991 International Symposium on Circuits and Systems (ISCAS'91)*, 5, 2573–2576, Singapore, 1991.

[63] P. Dudek, S. Szczepanski, and J. Hatfield, A high-resolution CMOS time-to-digital converter utilizing a Vernier delay line. *IEEE Transactions Solid-State Circuits*, 35, 240–247, Feburary 2000.

[64] P.M. Levine and G.W. Roberts, A High Resolution Flash Time-to-Digital Converter and Calibration for System-on-Chip Testing. *IEE Proceeding-Computers and Digital Techniques*, 152(3), 415–426, May 2005.

[65] R. Mullins and S. Moore, Demystifying Data-Driven and Pausible Clocking Schemes. *Proc. 13th Intl. Symp. on Advanced Research in Asynchronous Circuits and Systems (ASYNC)*, 2007, pp 175–185.

[66] S. Das, S. Pant, D. Roberts, S. Lee, D. Blaauw, T. Austin, T. Mudge, and K. Flautner, A self-tuning DVS processor using Delay-error Ddetection and correction. *Digest of Technical Papers, 2005 Symposium on VLSI Circuits*, pp 258–261.

[67] D. Bormann, GALS test chip on 130 nm process. *Electro Nobes Theor. Comput. Sci.*, 146(2), 29–40, 2006.

[68] K. Yun and A. Dooply, Pausible clocking based heterogeneous systems. *IEEE Transactions VLSI Systems*, 7(4), 482–487, December 1999.

[69] J. Sparsø and S. Furber, *Principles of Asynchronous Circuit Design–A Systems Perspective*, Kluwer Academic Publishers, 2001.

[70] I. Sutherland, Micropipelines: *Turing Award Lecture.Communications of the ACM*, 32(6), 720–738, June 1989.

[71] J. Kessels, A. Peelers, P. Wielage, and S.-J. Kim, Clock synchronization through handshaking. *8th Intl. Symp. on Advanced Research in Asynchronous Circuits and Systems (ASYNC)*, 2002, pp 59–68.

[72] M. Krstic, E. Grass, and C. Stahl, Request-Driven GALS Technique for Wireless Communication System. *11th Intl. Symp. on Advanced Research in Asynchronous Circuits and Systems (ASYNC)*, 2005, pp 76–85.

[73] W. Lim, Design methodology for stoppable clock systems. *IEE Proc. Computers and Digital Techniques*, 133(pt.E)(l), 65–69, January 1986.

[74] S. W. Moore, G. S. Taylor, P. Cunningham, R. D. Mullins, and P. Robinson, Self-calibrating clocks for globally asynchronous locally synchronous systems. *Proc. Intl. Conf. on Computer Design (ICCD)*, 2000, pp 74–79.

[75] R. Dobkin, R. Ginosar, and C. P. Sotiriou, Data synchronization issues in GALS SoCs. *10th Intl. Symp. on Advanced Research in Asynchronous Circuits and Systems (ASYNC)*, 2004, pp 170–179.

[76] A. E. Sjogren and C. J. Myers, Interfacing synchronous and asynchronous modules within a high-speed pipeline. *Transactions on Very Large Scale Integration (VLSI) Systems, IEEE*, 8(5), 573–583, October 2000.

[77] D. Bormann and P. Cheung, Asynchronous wrapper for heterogeneous systems. *Proc. Intl. Conf. on Computer Design (ICCD)*, pp 307–314, 1997.

[78] C.A. Petri, *Kommunikation mit Automaten*. PhD Thesis, Bonn, Institut für Instrumentelle Mathematik, 1962.

[79] T. Murata, Petri Nets: Properties, analysis and applications. *Proc. IEEE*, 77(4), 541–580, April 1989.

[80] L.Ya. Rosenblum and A.V. Yakovlev, Signal graphs: from self-timed to timed ones. *Proc. Int. Workshop on Timed Petri Nets*, Torino, Italy, July 1985, IEEE CS Press, pp 199–207, 1985.

[81] T.-A. Chu, C.K.C. Leung, and T.S. Wanuga, A design methodology for concurrent VLSI systems. *Proc Int. Conf. Computer Design (ICCD)*, pp. 407–410. IEEE CS Press, 1985.

[82] A. Yakovlev, A. Petrov, and L. Lavagno. A low latency asynchronous arbitration circuit. *IEEE Trans. on VLSI Systems*, 2(3), 372–377, September 1994.

[83] A. Yakovlev, Designing arbiters using Petri nets. *Proc. Workshop on Asynchronous VLSI*, Nof Genossar, Israel, March 1995, VLSI Systems Research Center, Technion, Haifa, Israel, pp 178–201.

[84] M. Renaudin and A. Yakovlev, From Hardware Processes to Asynchronous Circuits via Petri nets: an Application to Arbiter Design. *Int. Workshop on Token-Based Computing (ToBaCo'04) within ICATPN'04*, Bologna, Italy, pp 59–66, June 2004.

[85] J. Cortadella, L. Lavagano, P. Vanbekbergen and A. Yakovlev, Designing asynchronous circuits from behavioural specifications with internal conflict. *Proc. Int. Symp. on Advanced Research in Asynchronous Circuits and Systems (ASYNC'94)*, Salt Lake City, Utah, pp 106–115, IEEE CS Press, Nov. 1994.

[86] J.-B. Rigaud, J. Quartana, L. Fesquet and M. Renaudin, High-level modeling and design of asynchronous arbiters for on-chip communication systems. *Proc. Design, Automation and Test in Europe (DATE'02)*, p 1090, March 2002.

[87] Bystrov, D.J. Kinniment, and A. Yakovlev, Priority arbiters. *Proc. ASYNC'00*, pp 128–137. IEEE CS Press, April 2000.

[88] Mitrani and A. Yakovlev, Tree Arbiter with Nearest-Neighbour Scheduling. *Proc. 13th Int. Symp. on Computer and Information Sciences (ISCIS'98)*, 26–28 October, Belek-Anatlya, Turkey. In: *Advances in Computer and Information Sciences'98* (Eds. U. Gudukbay, T. Dayar, A. Gursoy, E. Gelenbe) Concurrent Systems Engineering Series Vol. 53, pp 83–92.

[89] K.S. Low and A. Yakovlev. *Token Ring Arbiters: an Exercise in Asynchronous Logic Design with Petri Nets, TR. no. 537*, Dept of Comp. Sci., University of Newcastle upon Tyne, November 1995.

[90] B. Grahlmann, The PEP Tool. *LNCS 1254: CAV'97 (Computer Aided Verification)*, Haifa, pp 440–443. Springer-Verlag, June 1997.

[91] V. Khomenko and M. Koutny, Towards an Efficient Algorithm for Unfolding Petri Nets, *LNCS 2154: CONCUR 2001 – Concurrency Theory*, pp 366–380, Springer Verlag, 2001.

[92] Benko and J. Ebergen, Delay-insensitive solutions to the committee problem. *Proc. ASYNC'94*, IEEE CS Press, pp 228–237, November 1994.

[93] Ch. E. Dickson, A Macromodule User's Manual. *Technical Report No. 25*, Computer Systems Laboratory, Washington University, St. Louis, Missouri, 1974.

[94] R.M. Keller, Towards a theory of universal speed-independent modules. *IEEE Trans. Computers*, C-23(1), 21–33, January 1974.

[95] C.H. van Berkel and C.E. Molnar, Beware the 3-Way Arbiter, *IEEE Journal of Solid-State Circuits*, 34, 840–848, 1999.

[96] O. Maevsky, D.J. Kinniment, A.Yakovlev, and A. Bystrov, Analysis of the oscillation problem in tri-flops. *Proc. ISCAS'02*, Scottsdale, Arizona, IEEE, vol. I, pp 381–384, May 2002.

[97] M.B. Josephs and J. Yantchev, CMOS Design of the Tree Arbiter Element. *IEEE Trans. VLSI Systems*, 4(4), 472–476, 1996.

[98] A.J. Martin, Synthesis of asynchronous VLSI circuits. In J. Straunstrup (ed., *Formal Methods for VLSI Design*, Chapter 6, pp 237–283. North-Holland, 1990.

[99] N. Thorne, On-chip buses enable block based ASIC/FPGA design. *IP'97-Europe*, October 1997, Bracknell.

[100] Clements, *Microprocessor System Design*, 3rd edn, Int. Thomson Publishers, 1997.

[101] D. Del Corso, H. Kirrmann, and J.D. Nicoud, *Microcomputer Buses and Links*. Academic Press, 1986.

[102] A.J. Martin, Collected Papers on Asynchronous VLSI Design, *Caltech-CS-TR-90-09*, Dept. of Computer Science, Caltech, 1990.

[103] D.E. Muller and W.S. Bartky, A theory of asynchronous circuits. *Proc. Int. Symp. on the Theory of Switching*, April 1959, Harvard University Press, pp 204–243.

[104] Bystrov and A. Yakovlev, Ordered Arbiters. *IEE Electronics Letters*, 35(11), 877–879, 27 May 1999.

[105] A.J. Martin, Collected Papers on Asynchronous VLSI Design, *Caltech-CS-TR-90-09*, Dept. of Computer Science, California Institute of Technology, 1990, p 50.

[106] D.J. Kinniment and J.V. Woods, Synchronization and arbitration circuits in digital systems. *Proc. IEE*, 123(10) 961–966, October 1976.

[107] J. Cortadella, M. Kishinevsky, A. Kondratyev, L. Lavagno, and A. Yakovlev, *Logic Synthesis of Asynchronous Controllers and Interfaces. Springer Series in Advanced Microelectronics*, vol. 8, Springer, 2002.

[108] A. Yakovlev, L. Lavagno, and A. Sangiovanni-Vincentelli, A unified signal transition graph model for asynchronous control circuit synthesis. *Formal Methods in System Design*, vol. 9, No. 3, Kluwer pp 139–188, November 1996.

Index